THE COSMIC CORE

PLANETS, STARS AND GALAXIES

AS THE UNIVERSE GETS BIGGER, THE GALAXYS GROW SMALLER

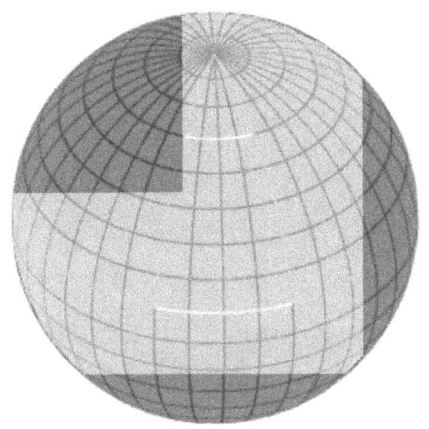

THE ONLY EXISTING QUANTIZED THEORETICAL MODEL

FOR FASTER THAN LIGHT SPACE TRAVEL

Made for the serious scientist by

RODNEY KAWECKI

2021 CY

We Hold These Truths to be Self Evident

Only for Geniuses

The Cosmic Core

Over time the universe slowly reveals itself. It is my hope that by reading this book you will comprehend factors in this novel that explain the product of a new gravity force theory. It's meaningless to keep believing an assumption when that assumption has been shown to be wrong with no shadow to the doubt. The earliest and most accepted theory in physics explains the analogy of time travel as a common hypothesis in advance physics. Yet the problem with this analogy lays with the advancement of intelligence and space travel with the idea that warp drive which is a faster than light technology is necessary to journey amongst the heavens. With this in mind this ability to travel faster than light eliminates any notion of time travel in the future.

As it stands today, the theory of gravitation has been unmeasured and assumed to be a force of attraction, abided basically by assumption of the illustrators and editors of its history. Really no theorists whom evolved in the scientific community has physically been shown that the gravity force has been defined positively as an attraction force but the historic direction of all that research seems to point to that. The positive and negative virtues from Galileo, Isaac Newton to Albert Einstein illustrate shortcomings when it comes to admitting it. In Newton's era it was a train of thought that secrets of scientific information were to be disguised in the diagrams and mathematics of their inventor so to keep secret the true nature of the illustrator's knowledge. Isaac Newton was one of these scientists. The public universities and colleges teach the gravity force as an attraction force even when the evidence shows likewise.

It's possible that the information was not available at the time of the research or the researcher was looking for something else when he wrote it. But evidence that is directly defined by the scientific inventor whether here and now postulates the positive labeling in the theory of gravity. The fact that 'similar masses" or stuff repels is a direct evidence to the contrary evidence of the historical claims that says it is not it is the truth and magnitude in the definition that the arbitrary gravity force theory stands by and is condemned to even today.

$$p=mv$$

Rodney Kawecki studied cosmology for most of his life. It wasn't until 1997 that he published his first book in advance physics. He lives in Los Angeles California in the San Fernando Valley.

As a child young Rodney and his family lived there for most of their lives. His study about gravitation and space exploration about the universe stems from insight to what people might call the evolution of the science when as an early young youth like many children he reviewed basic observations about physics he discovered in the curriculum.

I hold great appreciation to Professor Carl Sagan whom I share the great dream of enlightenment with exploring the Legend of the Species and Exploration of the Universe with great detail. His great insight about the fourth-dimension and curve space is a great analogy that expands beyond any and all boundaries that may travel beyond the enlightenment of Erwin Hubble's invisible yet expanding two-dimensional universe and or sphere of the flatlanders.

TABLE OF CONTENT

Chapter One
The Logic about Time

Chapter Two
Repulsive Gravity Theory

Chapter Three
The Repulse Expanse Theory

Chapter Four
Can gravity be reversed

Chapter Five
Does The Universe Have Mass

Chapter Six
Can We Reverse Gravity

INTRO

WHAT UNEXPLAINED PHENOMENA EXITS AT THE CENTER OF THE UNIVERSE

The Cosmic Background Imagery is slowly coming to life,

Few scientists believe that there exists a center for the universe and its space field is an absolute flat single two-dimensional flatland as it is described by Albert Einstein in his literatures both special and general relativity theories. In the older models like in the theory of relativity published in 1905, it was believed that the universe was flat. It wasn't until some ten years after relativity theory was published that new insight to the universe's model began to change. Erwin Hubble in 1925 discovered that the universe was expanding. He also was able to account about the speed it was expanding through the photographs through his observations through his telescopes. It was at this time that a new model began to be transformed.

Hubble's observation did exactly that when he published his findings about the expansion of the universe in his dialogues. He described uniformity and acceleration of space between the galaxies and that the universe was expanding all right but it wouldn't exactly be until 2015 that the scientific community would start taking notice that his observations would be observed seriously. The entanglement between Einstein's earlier published theories ten years earlier were already in the common household environment which put the Hubble observations behind the scene.

Today, due to Hubble's observations a step further through the threshold forwards in the history of the cosmogony illustrated towards what today is known as the 'cosmic background imagery'. It is a graphical image of what some scientists believe the universe may well look like as a whole. Furthermore, it starts a new computerized graphical tale about the universe that is leading modern time scientist to that due to this expansion and uniformity of the galaxies that the universe should well be an observable concrete three-dimensional image of the actual universe as the cosmic background imagery sustains in this modern day analogy in physics. But beware it is not the universe that expands but rather it is the space fieldre itself that in it accelerates the uniformity throughout the galactic regions.

So in what is the further reality what unexplained phenomena exists at the center of the universe that excites this phenomena momentum and why? The question about the universe's center has puzzled most physicists and scientist around the world for many a decade it is the idea that the universe has a center at all that excites the observations. It was in 1905 when the first physicists tried to answer this question that physicist was Albert Einstein. He put a cap on the universe that lasted for over a hundred years. His claim was that the universe was flat that its center was an accumulation of space-time matter that if a ship were to travel in a straight line and in a long enough time in a straight west worthy direction it would end up in the same place it started from.

This idea was startling because for this travel coordinates to happen the universe would have to be round it was a contradiction of his own theories. Over the years and with this idea it was discovered in 1921 by the observations of a second scientist Edwin Hubble that he discovered that the universe was expanding. The galaxies were traveling away from one another and at a velocity that measured faster than light. In Einstein's view this was impossible because he had quantized the universe with an oval crystal ball at the speed of only light traveled but with the evidence that was only factually based this meant it was only theoretically connected light and matter mass and energy. The idea that one could travel in a straight line at a starting point somewhere in space and end up in the same place after the long journey seemed impossible as well for him with a two-dimensional flatlanderish universe theory and the idea that gravity was a natural effect between a space and matter conflict in his space field equations would follow him for all the rest of days of his life were drawn in by the gates of time as Hubble's observations opened the doorway to not only universe expansion but a round three-dimensional universe hypothesis.

Space was never a dictum in Hubble's view though. It seemed to him observable evident if the galaxies were spreading a part from one another what some believed as the runaway universe was just that a runaway universe. His final discovery was that the universe and that everything he observed and seen as a planetary matter through his telescopes existed was the invisible activity of an expanding universe that can be orchestrated like a balloon that was expanding from somewhere underneath the space was invisible to the naked eye and was the three-dimensional shape of a common household inflating balloon as he called it.

So, what would one say about a round expanding universe as a source that deemed to be the foundation of this magic surface claimed to be its gravity force itself in an invisible crusade with the shape and inflationary model of a balloon for the invisible source of gravitation that due to a zero point energy field surrounding this presents could well be a gravity force without energy. Like Earth's gravity force it takes an attraction force and energy to form a gravitation environment. With that this balloon type bubble universe restrained by zero energy in a solid vacuum but is still a conditional force. An expanse force so great it causes the galaxies to remain steady state clinging to its expanding surface and three-dimensional space.

Space of course exists as a vacuum. It retains no measurable quantized energy except from that that exists in its suns and planetary bodies that exist in it. The invisible force of gravitation in space emptiness except for a pushing force that acts as a carrier force for galactic bodies throughout the realm is a force that increases the acceleration of momentum of anything that can gain new available speed from its field into a set velocity that increases and excels the instruments of velocity far beyond normal measurements. Like earth's gravity force an automobile accelerated to 50 miles per hour – takes 100 miles per hour of energy propulsion {p} to achieve. In space we have the same measurements except that this pushing force accelerates the normal rotation and orbital forces to exceed them to twice the acceleration instantaneously. A ship traveling through space at 50,000 miles an hour travels 99.9999 percent that acceleration plus another 50,000 miles an hour due to a new natural quantized expansion push force make sense?

This new theory in modern physics opens a unique pathway about the theoretic in physics that have yet been known to us till now. Under the new label ""The Quanta Physics Theory "" it's been a long 20 year study stemming from the ideas shown in relativity of both SRT and GRT. The imaginary light journey tour of special relativity and the equation $E=mc^2$, Rodney Kawecki's claim that $E=mc^2$ may be essential when talking about matter but in space the equation does not do very well. The space field has special characteristics that were never observed in 1905 or in 1915 and only come into view at the mythical time of Erwin Hubble's discoveries in 1924 ten years later where time is not considered absolute as in the decades before him.

Though Hubble's Universe only advances physics to an expanding universe Rodney Kawecki has re-ventured beyond the idea of a round inflationary universe with new equations that rebuild the universe in the three-dimensional model. His work in modern physics has ventured since 1997 when he published his first book "The Supertellic Electro-Magnetic Gravitational Universe Technology Theory" short for "SEGUTT" because of this a ship traveling in a straight line on a flat surfaced universe doesn't fall off at its edge but the ship returns to its place of origin as can be easily viewed in Einstein's allegations of 1915. But other changes occur like also the age of the universe, the length of it as well as how fast a ship can travel in a zero point space field push force gravitation universe field.

Specific rights lead modern physics into a foundation that is of uncertainty and is unknown today by modern science with the observations made by earlier scientific predictions that seem over the years have led us to believe the likewise of them 'If this is true than we stand at the threshold to regain a further gravity well of knowledge that can fill those mishaps'.

Rodney Kawecki is a modern physics writer of over fifteen new *ebooks* about the invisible forces that lay dormant throughout the grant rage only to be generated by the close calls of external interior reference frames. Written under the name " The Quanta Physics Theory " Rodney Kawecki believes he owes it to science that his observations in his books come to an end that only can be viewed as unique and remarkable. That in the aftermath he believes they might be what opens the doorway to vast speeds under limits beyond measurable links between light speed and velocities that can be achieved with a greater understanding about our universe though the bits and pieces written by modern physics authors that are only small stories about a factual universe beyond that we believe to be a perfect science. These shorter stories bring out great ideas about a universe and the foundations never completely filled into the history of how our universe works. It is by this vantage point that Kawecki's hopeful dreams that they do.

When we speak of today's physics we talk about a 93,000 billion light year length in a two-dimensional flatlander universe and age of a thirteen billion year old universe by its age in what is known as a cosmological speed limit of a crystal ball. The specific's designed in a round type "B' Hubble universe remake of the construction of the modern universe that change all the standing equations said in this paragraph but for some godly reason never made the mathematics' table beyond the discoveries of Hubble's round balloon, the universal depths of the universe and push of the gravity.

CHAPTER ONE

The Logic about Time

We all realize at least most of us interested in advancing physics that the upgrade of advance physics as it is today started out by being a science fiction story about a time traveler that could journey through space and time. In 1914 five years later, Albert Einstein also published another paper called "The General Theory of Relativity 'that went on to make his claims in special relativity his first book an actual reality. The real question though is if those definitions and equations were found to be mistakes later in Einstein's future why are they still part of the serious curriculum?

In this book you will read about obstructions or errors in his stories and other theories that over time have become evident. Some physicists believe that it is a right that errors become novice written by physicists and theorists and those that it is only nature for this to happen after all nothing is perfect.

The universe is expanding. It's retaining more and more area space is causing the galaxies to spread a part at their seams. How long will happen? Forever some believe? There is a limit some say. One of these great scientists of the age was Albert Einstein. He believed the universe was governed by the velocity of light and like a bubble everything about it was inside it. That light redeemed everything we need to know about the universe. Though the speed of light is very fast - the universe is more than that. Besides his universe at the end of his ride was a big event - we now know we can travel faster than light due to what's called "the reverse-vacuum expansion constant. That constant was discovered by Erwin Hubble in the year 1928. He was the one physicist that discovered the universe was expanding and measured how fast. Besides Albert Einstein he was the one whom open a new doorway to the cosmos the one physicists use today.

But what is a reverse vacuum. It's a time in the duration of the big bang event that the warp age and expansion of a primeval egg exploded and stretched across the cosmos from where it resided outwards to the maximum limit of its existence. As all matter stretched across empty space the fabric of the space it evolved from like a rubber band it stretched outwards and stretched back again from point A back to point B amidst an empty dark space. Space is an element that is not transference unless observed by moving objects. Everything in space is formally at rest- it is the momentum of the gravity of space that creates momentum amongst the planets, stars and galaxies and all the stars and planets are inside galaxies.

It was a reverse vacuum activity that allowed an empty void to be opened in the middle of that action. When all matter particles and subatomic masses reformed back into gigantic chunks of matter the bang had opened a briefcase at the middle of the space that formed into a concentrated bubble of pressure the same that like today is what expands the universe. Erwin Hubble explains this as a balloon that is slowly expanding itself and it seem it is indefinitely. How and why what is it that is written in this great novel about space.

Reverse Vacuum Space Concentrate

A concentrate is a form of substance which has had the majority of its base component (in the case of a liquid: the solvent) removed. Typically, this will be the removal of water from a solution or suspension, such as the removal of water from fruit juice. One benefit of producing a concentrate is that of a reduction in weight and volume for transportation, as the concentrate can be reconstituted at the time of usage by the addition of the solvent.

It will also allow the enlarging of the mass concentrate to increase if it subtend with a continuous flow of cold liquid such as water igniting its growth or expansion causing the solvent to multiple. The concept of expiration is related but legally distinct in some jurisdictions. The removal of water from a solution or suspension such as the removal of water from a naked fruit of its juice one benefit of producing a concentrate is that of space pressure reverse vacuum expanse.

Shelf-life depends on the degradation mechanism of the specific product. Most can be influenced by several factors: exposure to light, heat, and moisture, transmission of gases, mechanical stresses, and contamination by things such as micro-organisms. The availability of a constant and abundance of supplies product quality is often mathematically modeled around a parameter (concentration of a chemical compound, a microbiological index, or moisture content).The universe existing of matter is only 5 % percent. It will keep its best quality for 12 to 49 billion years to mature.

In chemistry, a suspension is a heterogeneous mixture that contains solid particles sufficiently large for sedimentation. The particles may be visible to the naked eye, usually must be larger than one micrometer, and will eventually settle, although the mixture is only classified as a suspension when and while the particles have not settled out. A suspension is a heterogeneous mixture in which the solute particles do not dissolve, but get suspended throughout the bulk of the solvent, left floating around freely in the medium. [1] The internal phase (solid) is dispersed throughout the external phase (fluid) through mechanical agitation, with the use of certain recipients or suspending agents. An example of a suspension would be sand in water. The suspended particles are visible under a microscope and will settle over time if left undisturbed.

This distinguishes a suspension from a colloid, in which the suspended particles are smaller and do not settle. [2] Colloids and suspensions are different from solution, in which the dissolved substance (solute) does not exist as a solid, and solvent and solute are homogeneously mixed.

A suspension of liquid droplets or fine solid particles in a gas is called an aerosol. In the atmosphere, the suspended particles are called particulates and consist of fine dust and soot particles, sea salt, biogenic and volcanogenic sulfates, nitrates, and cloud droplets.

Suspensions are classified on the basis of the dispersed phase and the dispersion medium, where the former is essentially solid while the latter may either be a solid, a liquid, or a gas.

Water is a transparent, tasteless, odorless, and nearly colorless chemical substance, which is the main constituent of Earth's hydrosphere, and the fluids of most living organisms. It is vital for all known forms of life, even though it provides no calories or organic nutrients. Its chemical formula is H_2O, meaning that each of its molecules contains one oxygen and two hydrogen atoms, connected by covalent bonds. Water is the name of the liquid state of H_2O at standard ambient temperature and pressure. It forms precipitation in the form of rain and aerosols in the form of fog. Clouds are formed from suspended droplets of water and ice, its solid state when finely divided, crystalline ice may precipitate in the form of snow. The gaseous state of water is steam or water vapor. Water moves continually through the water cycle of evaporation, transpiration (evapo transpiration), condensation, precipitation, and runoff, usually reaching the space sea.

Space resembles a dark blue radiated from darkness is the affect between visible light and cold moister in the space element

The word water comes from Old English wæter, from Proto-Germanic *watar (source also of Old Saxon watar, Old Frisian wetir, Dutch water, Old High German wazzar, German Wasser, Old Norse vatn, Gothic wato), from Proto-Indo-European *wodor, suffixed form of root *wed- ("water"; "wet"). Also cognate, through the Indo-European root, with Greek ύδωρ (ýdor), Russian вода́ (vodá), Irish uisce, Albanian ujë.

Cosmology deals with the big questions of the universe, often the same questions that keep philosophers up at night. When did the universe begin? How did it start? Has the universe always been expanding? (For the record, the answers are: about 13.8 billion years ago, in a high-density state that rapidly expanded called the Big Bang and yes, but not always at the same speed.) But here's a question they haven't figured out yet: How's it all going to end?

It's a big question all right, but we've made surprising headway toward an answer. In the last years of the 20th century, the astrophysical community was stunned to learn that the universe was driving itself apart. For decades, scientists had known that distant galaxies all move away from us, with the farther ones moving the fastest. The only way this makes sense is if the universe itself is expanding. Given all the matter in the cosmos, the force of gravity should be slowing down that expansion. But when cosmologists calculated just how much it's slowed down, they get a negative result — the expansion of the universe is speeding up!

Nobody knows what's driving the acceleration, so cosmologists have dubbed that mystery dark energy. It is so dominant (about 69 percent the total content of the entire cosmos) that dark energy quickly became a part of any discussions about the final end of the universe. And while there are no definite answers yet, those discussions have come up with a few interesting possibilities.

Until recently, asking what happened before the Big Bang was generally considered by physicists to be a religious question. General Relativity Theory just does not go there. As time goes to zero, General Relativity spews out zeros and infinities. So the question did not make sense from a mathematical/scientific point of view.

Thanks to studies on the rate the universe is expanding, and applying this knowledge in reverse, we now know the universe is roughly 13.8 billion years old. Everything we can observe in our solar system, other galaxies and everything in between, Big Bang Theory says inflated out extremely rapidly from an initial point much smaller than an atom. The Big Bang Model is currently our best explanation for why the cosmos appears as it does. However, the Big Bang Model is not able to answer some of the more challenging questions, including - what preceded it and what caused it.

This reasoning applies to the Universe as a whole. Since Edwin Hubble's observations in the 1920s, we've known that the galaxies are moving apart from one another: the Universe is expanding. Traditionally we have argued that the ultimate determinant of the galaxies' future is the amount of mass contained in and around galaxies. If this mass is large enough, its gravity should be strong enough to halt the outward movement of galaxies, causing them to eventually collapse together in a reverse of the Big Bang, a so-called Big Crunch.

Dust along the plains of the Milky Way

NASA

Will the Universe end in a fiery Big Crunch and will this be the real Armageddon?

The new observation of an apparently accelerating Universe implies that some kind of cosmic 'antigravity' is at work. This may sound like the stuff of science fiction, but it has a basis in physics. If you endow empty space with energy, the gravitational effect of this energy has the strange property of producing a new repulsive force throughout all of space. While this theoretical insight has been known for over a generation, the common wisdom was that such energy in empty space must be precisely zero.

In fact, when Einstein first laid out the equations that govern the large-scale structure of the Universe in 1916, he introduced the possibility of universal repulsion in the shape of a quantity called the cosmological constant - without knowing its significance on the microscopic scale. However, he quickly dispensed with the idea, calling the cosmological constant his "biggest blunder".

You might imagine that if the expansion of the Universe is accelerating, this implies that the Universe will go on expanding forever. But things are not that simple. We don't know the source of the energy of the vacuum and so it may have properties we are currently unaware of. Here are three possibilities:

The vacuum energy decreases with time. The acceleration also slowly decreases, and ultimately the future of the Universe will once again be determined by the gravitational attraction of the matter within it.

The present inferred acceleration is later proved to be incorrect, but there is still a tiny amount of vacuum energy. You might suspect that once again an inventory of all the matter in the Universe should allow us to determine the ultimate long-term behavior of the expansion. Alas, this is not the case. If we found there was enough matter for us to be heading towards a Big Crunch, the energy in the vacuum can have unexpected effects. Suppose that empty space possesses an amount of energy only one-thousandth the amount needed to measurably affect the present expansion. This energy would still eventually forestall the ultimate collapse of a universe in which matter currently appears to have the upper hand.

Finally, and perhaps most unusual of all, once we realize that empty space has energy, nothing forbids this energy from being negative. If this is the case, then even an unimaginably small negative energy in empty space will ultimately cause the Universe to re-collapse, independent of how much matter now exists therein.

Reverse vacuum gravity is a review about space that allows a concentrated mass, like the balloon hypothesis about an expanding universe to supersede itself. Unlike a false vacuum the fate of the universe is not observed in the final hours. It slowly redeems itself backwards until the universe falls again a time when all the forces we believe hold the universe diminish and a collection at the bottom of the most gigantic gravity well emerges.

How old is our universe? How did it begin? How big is it and what will happen?

Age of Universe is measured by the age of the galaxies not planets becuase the galaxies are older

The Big Bang Theory is the leading explanation about how the universe began. At its simplest, it says the universe as we know it started with a small singularity, then inflated over the next 13.8 billion years to the cosmos that we know today.

Because current instruments don't allow astronomers to peer back at the universe's birth, much of what we understand about the Big Bang Theory comes from mathematical formulas and models.

Called the Methuselah Star, HD 140283 is 190.1 light-years away. Astronomers refined the star's age to about 14.5 billion years (which is older than the universe), plus or minus 800 million years.

In the Quanta Physics study and book you are now reading. It has come to the table that the Methuselah star is not older than the universe.

In The Quanta Physics Theory, Rodney Kawecki explains a gap in the time coordinates of the aging of time and space physicists use in measuring age data throughout the cosmos. In theory, he recognizes that just after the big bang event the reformation of matter re-blemished back into mass show the chunks of matter immediately reforming into galaxies throughout the cosmos system like they are today but at the earliest time of creation. These gigantic chunks of reinvented matter spawned into separated deities were young enough that their energies were close to if not almost equal with the earliest mass they were spawn from the big bang. Small as they became the energy in their masses were enough to cause a second stage of evolution.

What this means is the energetic content of the earliest galaxies energy volume was enough that in the midst of re-formation over a specific length of time they later exploded leaving behind erotic black hole formations at the midst of them all.

When scientists explain the origin of reformation after the big bang there is not recollect of stages that may have occurred. In The Popcorn Expanding Universe Rodney Kawecki has published a 250 page reconstruction of what might be loss information that has led scientist to discover older planetary model's in space but with no explanation why they exist with such a greater age volume than is known to them.

But in a universe and the big bang as it suggest expanding at a vast velocity as it is and without the counter play to the suggested energy-mass of a three-dimensional deity estimates might be at loss with no explanation. The length, the mass and the universe's age all come into play when Kawecki speaks about a greater expanding universe as he does in his books.

This is a backyard view of the sky surrounding the ancient star, cataloged as HD 140283, which lies 190.1 light-years from Earth.

Methuselah star

Though the expansion and loss of stress dark matter has as a repulsive force against the galaxies is the bundling of tension dark matter puts on galactic masses prove to show a relief of tension as the universe expands at its continuum. The loss of tension also relieves a planetary bodies raise or fall in energy readings leaving behind a false reading. How long matter has resided in space reluctant to later expansion readings that may vary over time has relieved us from what the universe's age might be over 10 billion years ago and now. This gap in detecting the energy volume from an earlier date reluctant to the Hubble Constant might read differently if it were not the distant between expansions that is observed by the energy volumes of the bodies using the same method.

For more than 100 years, astronomers have been observing curious stars located some 190 light years away from Earth in the constellation Libra. It rapidly journeys across the sky at 800,000 mph (1.3 million kilometers per hour). But more interesting than that, HD 140283 — or Methuselah as it's commonly known — is also one of the universes oldest known stars. Collaborators estimated HD 140283's age to be 14.46 billion years — a significant reduction on the 16 billion previously claimed. That was, however, still more than the age of the universe itself, but the scientists posed a residual uncertainty of 800 million years, which some said made the star's age compatible with the age of the universe.

It is about midfield these predictions that it has been assumed that aside in the universe's backyard the universe evolved through stages that may have yet been realized. Stage one the big bang. Stage two the smaller bangs of the galaxies. Evidence shows that there exist evidence of black holes that exist at nearby center of every one of the galaxies. Some of these black holes may have closed. But the ones that are open those that are of the Milky Way' Galaxy is evidence of an extreme explosion that may be evidence of a second stage evolution. The stage two smaller little bang theory which is expressed in Rodney Kawecki's earlier book called "The Popcorn Universe". It also illustrates why all the stars, planets and system of stars and solar systems are all formations inside the galaxy and is the only place they exist by nature.

It also goes on to be evidence Kawecki has related to the Universe Expanse Theory" that he illustrates to the gravity theory expressed in the nature of an ever-growing universe cosmic background imagery the place from which the expanse is coming from. He also illustrates how this expanse and growth that is expanding the universe in a nature state and constant rate of expansion is the source that illuminates the gravitation in space. He states that if not for the expanse the universe would be a closed vacuum and have no natural momentum or motion amongst it.

A higher value for the Hubble Constant indicates a shorter age for the universe. A constant of 67.74 km per second per mega parsec would lead to an age of 13.8 billion years, whereas one of 73, or even as high as 77 as some studies have shown, would indicate a universe age no greater than 12.7 billion years. It's a mismatch that suggests, once again, that HD 140283 is older than the universe.

Conclusions are based on the idea of an expanding universe, as shown in 1929 by Edwin Hubble. This is fundamental to the Big Bang — the understanding that there was once a state of hot denseness that exploded out, stretching space.

Bigger Unanswered Questions about Dark Matter

A higher value for the Hubble Constant indicates a shorter age for the universe. A constant of 67.74 km per second per mega parsec would lead to an age of 13.8 billion years, whereas one of 73, or even as high as 77 as some studies have shown, would indicate a universe age no greater than 12.7 billion years. It's a mismatch that suggests, once again, that HD 140283 is older than the universe. It has also since been superseded by a 2019 study published in the journal Science that proposed a Hubble Constant of 82.4 — suggesting that the universe's age is only 11.4 billion years.

Reluctantly, we are advancing technology to greater propulsion level's necessary to meet our future goal in space knowledge. Our journey to another world some might call it. Over the past 20 years we have advanced is technology for space travel here on earth to venture the stars.

Superfast Military Aircraft

The Darpaglider, called the Falcon Hypersonic Test Vehicle 2 (HTV-2), blasted off from Vandenberg Air Force Base in California atop a Minotaur 4 rocket.

The HTV-2 vehicle was expected to reach suborbital space, then re-enter Earth's atmosphere and glide at hypersonic speed to demonstrate controllable flight at velocities of around Mach 20, which is about 13,000 mph. At that speed, more than 20 times the speed of sound, a vehicle could fly from New York City to Los Angeles in 12 minutes.

The speed of light is equal to exactly 299,792,458 meters per second, or 670,616,629 miles per hour.

CHAPTER TWO

Repulsive Gravity Theory

This theory holds that it is the altered shape of space, deformed by massive objects, that causes gravity, which is actually a property of deformed space rather than being a true force. Both general relativity and Newtonian gravity appear to predict that negative mass would produce a repulsive gravitational field.

Newton's famous second law, which describes how apples fall from trees and satellites stay in orbit, can be derived from these underlying microscopic building blocks. Extending his previous work and work done by others, Verlinde shows how to understand the curious behavior of stars in galaxies without adding the puzzling dark matter.

The outer regions of galaxies, like our own Milky Way, rotate much faster around the center than can be accounted for by the quantity of ordinary matter like stars, planets and interstellar gasses. Something else has to produce the required amount of gravitational force, so physicists proposed the existence of dark matter. Dark matter seems to dominate our universe, comprising more than 80 percent of all matter. Hitherto, the alleged dark matter particles have never been observed, despite many efforts to detect them.

No need for dark matter

Accordingly there is no need to add a mysterious dark matter particle to the theory. In a new paper according to the holographic principle, all the information in the entire universe can be described on a giant imaginary sphere around it. This idea shows that this idea is not quite correct—part of the information in our universe is contained in space itself. This extra information is required to describe that other dark component of the universe: Dark energy, which is believed to be responsible for the accelerated expansion of the universe. Whereas ordinary gravity can be encoded using the information on the imaginary sphere around the universe, as he showed in his 2010 work, the result of the additional information in the bulk of space is a force that nicely matches that attributed to dark matter. The dark matter theory was first spoke about by Albert Einstein when he illustrated how dark matter, dark energy is the result of matter pushing space away from it. "Many theoretical physicists like me are working on a revision of the theory, and some major advancement has been made. We might be standing on the brink of a new scientific revolution that will radically change our views on the very nature of space, time and gravity." The nature of that future theory is in this book.

For 80 years, scientists have puzzled over the way galaxies and other cosmic structures appear to gravitate toward something they cannot see. This hypothetical "dark matter" seems to outweigh all visible matter by a startling ratio of five to one, suggesting that we barely know our own universe. Thousands of physicists are doggedly searching for these invisible particles.

But the dark matter hypothesis assumes scientists know how matter in the sky ought to move in the first place. Before Einstein, space seemed featureless and changeless, as Isaac Newton had defined it two centuries earlier. And time, Newton declared, flowed at its own pace, oblivious to the clocks that measured it. But Einstein looked at space and time and saw a single dynamic stage — space-time — on which matter and energy strutted, generating sound and fury, signifying gravity.

Newton's law of gravity had united the earthly physics of falling apples with the cosmic dances of planets and stars. But he couldn't explain how, and he famously refused to try. It took an Einstein to figure out gravity's true modus operandi. Gravity, Einstein showed, did not just make what goes up always come down. Gravity made the universe go 'round or did it.

Gravity's secrets succumbed to Einstein's general theory of relativity, unveiled in a series of papers submitted a century ago this November to the Prussian Academy in Berlin. A decade earlier, his special theory of relativity had merged matter with energy while implying the unity of space and time (soon to be christened as spacetime). After years of struggle, Einstein succeeded in showing that matter and spacetime mutually interact to mimic Newton's naïve idea that masses attract each other. Gravity, said Einstein, actually moved matter along the curving pathways embodied in spacetime — paths imprinted by mass and energy themselves. As expressed decades later by the physicist John Archibald Wheeler, mass grips spacetime, telling it how to curve, and spacetime grips mass, telling it how to move. Einstein's theory explained a famous observation that Newtonian gravity could not: a subtlety in the orbit of the planet Mercury. And his equations implied further slight deviations from Newtonian calculations. Over the last century, general relativity's predictions have been repeatedly verified by modern precision measurements to physicists today, general relativity and gravity.

But general relativity is about more than just understanding gravity. It's about explaining the totality of existence.

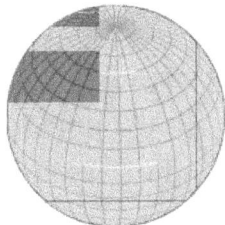

Cosmic Geodesic Background Image of the Universe

The curvature of spacetime lies at the heart of general relativity. The theory predicts that anything moving through a gravitational field undisturbed by other forces will follow a curved path called a geodesic. Geodesics on two-dimensional curved surfaces, like the Earth's, illustrate how curvature creates gravity. From any point on the equator, for instance, the shortest path to the North Pole follows a curve — the geodesic corresponding to a meridian. If two people start out on such a trek, starting some distance apart, they pursue different curved meridians to the pole but grow closer together as they travel northward. It would appear that the curvature was pulling them towards each other — just as Newton's gravity described.

Falling freely

Einstein's imagination gave birth to general relativity's core idea as he gazed out his office window while he was supposed to be evaluating patents. "All of a sudden he was struck by a thought," Einstein later said. "If a person falls freely, he will certainly not feel his own weight."

Falling freely is measured parallel with Newton's equation 9.8 m/s. that is how fast an object falls through the air by it. The only effect on that object is gravity and air of course. Depending on what force that gravity measures on which planet is the speed the object will fall.

A History of the search for Gravitational Waves

1915 - Albert Einstein publishes general theory of relativity, explains gravity as the warping of spacetime by mass or energy

1916 - Einstein predicts massive objects whirling in certain ways will cause spacetime ripples—gravitational waves

1936 - Einstein has second thoughts and argues in a manuscript that the waves don't exist—until reviewer points out a mistake

1962 - Russian physicists M. E. Gertsenshtein and V. I. Pustovoit publish paper sketch optical method for detecting gravitational waves—to no notice

1969 - Physicist Joseph Weber claims gravitational wave detection using massive aluminum cylinders—replication efforts fail

1972 - Rainer Weiss of the Massachusetts Institute of Technology (MIT) in Cambridge independently proposes optical method for detecting waves

1974 - Astronomers discover pulsar orbiting a neutron star that appears to be slowing down due to gravitational radiation—work that later earns them a Nobel Prize

1979 - National Science Foundation (NSF) funds California Institute of Technology in Pasadena and MIT to develop design for LIGO

1990 - NSF agrees to fund $250 million LIGO experiment

1992 - Sites in Washington and Louisiana selected for LIGO facilities; construction starts 2 years later

1995 - Construction starts on GEO600 gravitational wave detector in Germany, which partners with LIGO and starts taking data in 2002

1996 - Construction starts on VIRGO gravitational wave detector in Italy, which starts taking data in 2007

2002–2010 - Runs of initial LIGO—no detection of gravitational waves

2007 - LIGO and VIRGO teams agree to share data, forming a single global network of gravitational wave detectors

2010–2015 - $205 million upgrade of LIGO detectors

2015 - Advanced LIGO begins initial detection runs in September

2016 - On 11 February, NSF and LIGO team announce successful detection of gravitational waves.

The Universe has a Dark Mystery

We can consider the universes center of mass to act in the same way as a black hole. It observes energy from outside and republishes the universe with a dark mystery.

The universe's center mass acts in the same manner as the black hole it observes moister and maintains an interior vacuum system that stays neutral in its temperature while at the same time causes the mass to increase in size and mass.

If you can imagine it the whole universe all matter bundled in one place. When the big bang occurred the material mass broke into smaller pieces known as galaxies. Again, as a secondary stage the galaxies exploded and the stars and planetary systems formed inside them. Space imploded along with all the matter as the galaxies broke into smaller pieces inside them in the aftermath. Space became colder as the small suns bundled and only warmed the nearby planets and stars inside the galaxies. That's where all the stars and planets and systems of them are inside none exist outside the galaxies except great distances between them. Outside the galaxies the space is colder as the distance between the galaxies was further away from each other and in the earliest era the galaxies that began almost frozen as they were smaller pieces of the bang.

Over time the moister inside the galaxies warmed and melted the ice around the early planets systems and the seas formed. As time continued early life began on some of the planet and an evolution of animal and human species developed. The area in space between the planets and stars and the galaxies were young and well denser and over time they moved further away from that closeness until a point of terminal similarity occurred and orientation became subtle throughout the cosmos. From that time pass the terminal similarity of the masses a constant evolution of expansion began and towards the center of the universe like black holes that observe lethal energy throughout empty space moister throughout the time expansion has been being observed that serves to maintain a space equilibrium that through time changes in terminal mass, energy and climate.

THE COSMIC CORE

What people or the species does not know about the universe is that long ago in its beginning when the bang happen another similar mass was created in center of everything and this similarity finally began to show its existence as it started the expanse of the universe. It has been expanding ever since at a controllable rate and velocity. As it expands the distance between the galaxies seem to shorten but it's just that the expanse is getting bigger and bigger.

As it does it shadows the galaxies nearest it and expands the space fabric around it and them. The galaxy's spread further and further apart as this expanse gets bigger and bigger. It has been since the beginning of time and its part of the enclosed universe nobody knew about. Some call it the cosmic background image coming to life but it's just an expanse in the universe. It is what make everything planetary, stars and systems of planets push, pull and rotate with its constant expanse over all the galaxies abroad space.

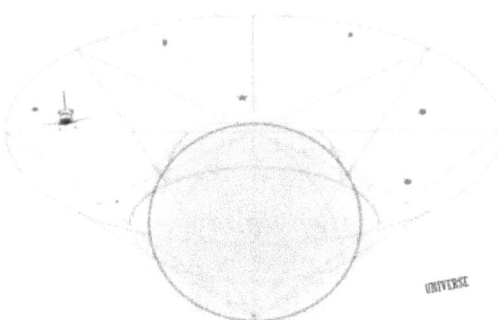

CHAPTER THREE
The Repulse Expanse Theory

Scientists discovered in 1998 that the Universe is expanding at an accelerating rate, the possibility that dark energy could explain the observation was intriguing. But because there has been little progress in figuring out exactly what dark energy is, the idea has since become more of a problem than a solution for some scientists. One physicist, Massimo Villata of the National Institute for Astrophysics (INAF) in PinoTorinese, Italy, describes dark energy as "embarrassing," saying that the concept is an ad hoc element to standard cosmology and is devoid of any physical meaning. Villata is one of many scientists who are looking for new explanations of the Universe's accelerating expansion that involve some form of repulsive gravity.

The gravitational repulsion between matter and antimatter is a prediction of general relativity. In this scenario, matter has a positive gravitational charge while antimatter has a (hypothetical) negative gravitational charge. As a result, both matter and antimatter are gravitationally self-attractive, yet mutually repulsive.

At the CernInstitute or Center for European Nuclear Research physicist Dragan Hajdukovic has been investigating what he *thinks* may be a widely overlooked part of the cosmos: the quantum vacuum. He suggests that the quantum vacuum has a gravitational charge stemming from the gravitational repulsion of virtual particles and antiparticles. Previously, he has theoretically shown that this repulsive gravity can explain several observations, including effects usually attributed to dark matter. Additionally, the additional gravity suggests that we live in a cyclic Universe (with no Big Bang) and may provide insight into the nature of black holes and an estimate of the neutrino mass. In his most recent paper, published in Astrophysics and Space Science, he shows that the quantum vacuum could explain one more observation: the Universe's accelerating expansion, without the need for dark energy.

In Quanta Physics Theory developed in 1997 by Rodney Kawecki, we perceive the universe as an expanding entity that in its inflation is pushing space that lays around it away from it in the same manner the galaxies are spreading a part away from each other. Yet at a constant speed is it a negative energy or a positive energy that is not the result of the action that is being shown to us. In reference to dark matter dark energy hypothesis dark matter might well be a dark energy that in response to discovering it like the dark bosom particles, might well be a positive repulse particle that is not only unseen able but acts in repel setting it on a different end of the vacuum. Dark gravity in space bundles and bunches up as the universe expands.

In it the universe is literally a spacious inflationary round ball that like a balloon inflates getting bigger and bigger and is big enough that it causes the rippling of the space about it to be physically curled around itself and make it act as a repulse force. Reluctantly, this mass at the center of the universe this balloon as Erwin Hubble calls it - explains the expanse of the matter universe and show the forces created by its action and a velocity that physically recede the velocity of light in the meaning.

It is not some energy or anti-quantum particle but is evidence of a physical inhabitance that's action is affecting not only the universe as a whole but the space around it a long with and the elastic space that recedes beyond that space. In Hubble's observation, we can view an external entity in the image of a gigantic ball that is so large that it is expanding the space between everything in the universe. Sucking in all the moister that lays with the dark fabric of space as an exchange for its expansion. We could also say that this expanding third element is in everything a particle membrane and exist with the space fabric element chemistry that is the actual the fabric of space that sits in between the particles and subatomic particles that energies bind together with because of this dark fabric that even the big bang event generated to bind them together at the time and era and in the bang that happen over thirteen billions of years ago.

In relativity, we assume the idea that it is energy that binds everything we observe - but lastly in the universes creation an entity that hides itself in clear view and causes a repulse wave as it acts as the universe's center of mass in the evolution of our universe we as scientist only observe the effects it embodies and shows to us by rippling the fabric of the space that when things come into a clear observation we are not puzzled by the idea that something physical might be missing in our physio-biologic theory about the universe.

We call the force of gravity a repulse expanse when it acts differently than what we observe to be normal. But we only observe the effects of that normal gravity force that a repulse action in such a theory must be the action of an anti-gravity force and without the evidence. Is this deity that is causing expanse in the universe the same and cause of what we view as anti-gravity in the motion of the space grid or can it be something much greater? That its action is the physical push force of an inflating bubble that acts like the inflation of a balloon being blown up with air that without cause of what it may actually be is of physical origin and the effects of what specific element, velocity, reverse engineering of the spacious dark gravity force because its expanse volume is so great and third dimensional that its mass is a lot more greater than what can be believed of it.

Only five persons knew what Albert Einstein was talking about when he published his book on Relativity. To those who didn't understand the theory he told "I'll explain it to you."

One quote: When an object falls it doesn't gain mass. It is cluttered by the senseless of having no force behind it. The reason for this is - it is falling with the 'downfall" rate of velocity. So, it isn't until the object reaches a point of regression equal with the downfall velocity that it does when that happens - the object starts gaining mass but not until. The object will glide with the earth's gravity constant as it falls but not beyond that until it reaches the ground.

Einstein believed that a moving object gained more and more energy when it was accelerated but logic concurs that a five pound object is still only a five pound object whether it called mass or an object.

In the same way a ship traveling through space with the freefall of the universal (repulsive) gravitation that it will reach the point of the freefall regression speed limit, meaning freefall has taken the object to the fastest regression point of its terminal velocity and the object will begin to gain mass but not until then.

Einstein's idea about space reluctantly illustrate mass verses velocity stigma but the fact of the matter an object or ship adrift space will not gain mass at all. Force needs to be applied and whereas Einstein's theory reassembles velocity and mass as a deterrent to acceleration it isn't. What is loss in space is an object mass. The ship wonders in midfield space adrift without acceleration in a space element that retains no source for a gravity force. Earths mass does not elevates weight as mass on a moving object because its gaining distance from the secondary object the earth itself. The further away it gets the less energy is repelled for its freedom to zero point energy - space. Any applied acceleration force is there only make the objects escape as it accelerates to escape the planets fit or pull on it. It is the objects rest mass that acquires how much force is needed for its journey to and foe' the area. As the object flees further away it advances less weight the higher it emerges.

In the cosmic journey at the atomic level mankind can only travel in the path that has already been built for them in the past.

In the cosmic journey at the cosmic level mankind can only travel towards the past from the path that has yet to be built in the future that they travel from and that has yet to be built by them.

Repulsive Gravity

"In the theory of general relativity, we usually assume that the energy is greater than zero, at all times and everywhere in the universe," says Prof. Daniel Grumiller from the Institute for Theoretical Physics at the TU Wien (Vienna). This has a very important consequence for gravity: Energy is linked to mass via the formula $E=mc^2$. Negative energy would therefore also mean negative mass. Positive masses attract each other, but with a negative mass, gravity could suddenly become a repulsive force.

Gravity is the effect to the off-spring of it's source. *Repulsive Gravity* has little to nothing to do with actual energy positive or negative effects of energy. Negative energy is the result of grounding a positive electric current. In repulsive gravity as a force the result of the force is negatively pushing something away from it, Positive masses attract each other, but with a negative mass, gravity they become a repulsive force.

Energy is linked to mass via the formula $E=mc^2$

Again, we have a situation where the unknown is labeled negative. If Hubble law if the universe's source were negative energy, the reverse of energy could not be grounded called positive grounding. The source would have to be positive to have a negative energy for the source for gravity. In Quanta Physics the source of repulsive gravity is a push force not energy as described as in an electric current. Positive repulsive force in the meaning is pushing away celeste planetary matter creating a force not trying to involve the repulsive force to any type of real energy.

You could say that the Hubble universe but according to quanta physics is a real force implied by the source as a imagery of that it is rippling its own space surface in a circular motion, allowing there to exist a naked vacuum that pushes against planetary objects causing them to spin on an axis And adding erotic spacious affects. It is nothing like borrowing positive energy from negative energy like in an electric current or an active electric affect that communizes a 50/50 current. Or is as A.E. explains in $E=mc^2$ where light compensates its energy in a fixed area force that acts in the same manner as electricity but is a radiant of quanta particles.

Energy cannot be obtained from nothing, even though it can become negative. Nor is any affect in a vacuum field be more than anything except accounted for as an interfering deity and of an external force. Quanta Physics makes no point that its repulsive force or even its energy is electrical. But that it's an effect on exterior objects that try to explain its source.

Space is a negative repulsive force that regulates as an undercurrent to streamline acceleration in the space field. Its source is repulsive as it is gaining area mass over the vacuum of space and therefore pushes as acceleration excels velocity. Any moving mass is physically accelerated by this under force like a roller scooter ride acts in mechanical free fall but a ship traveling through space is equipped with its own acceleration force as well and it is this type of action that warp drive occurs.

As the water is spilled it's instantaneously refilled by the liquid concentrated source. A source that has been condenses since the big bang event that created it. As the leakage of its space quantity escapes condensation the liquid mass remains whole causing instantaneous expansion of the grid plat formation.

Space expands, is flexible, bends, curves and warps in the presence of matter. It does all these things as the source that retains solidarity to the universe.

Universe space has nothing to hit but matter

Space is only affected by velocity or movement of an object. This means that if nothing moves sitting stationary in the dark. Space has no problem with it. But once it starts moving and depending on how fast it moves a vessel. Planet, star or galaxy gaining acceleration by some reason will not affect the space until its put in motion. As the object gains velocity space reacts by instantaneously protecting itself from change. Once an object gains a velocity close to light speed it has already pushed it away twice the optical velocity. As it still gains speed the faster it travels the more acceleration gain space adds to it transparent speed. Finally the object will gain curvature. I mean it will cause space to push it sideway against it fabric. When his happens it will enter acceleration free fall. This means that speed no longer accounts for its velocity - the mass now is traveling so fast its un-accountable because its traveling faster than what might be called normal acceleration. Since this free fall point is non-inertial the mass will slow down again until it reaches uniformity with the space fabric. It discovers a place in the cosmos with other vast deities around it space becomes subtle and the mass object is directed into a path relative to the subtle push away force of the fabric.

Imagine that - matter or objects don't gain their energetic life style until they have passed all the tests. The masses have a place in the cosmos but not until they have fallen.

Reverse vacuum gravity is a review about space that allows a concentrated mass, like the balloon hypothesis about the expanding universe to supersede itself. Unlike a false vacuum the fate of the universe is not observed in the final hours. It slowly redeems itself backwards until the universe falls again a time when all the forces we believe hold the universe diminish and a collection at the bottom of the most gigantic gravity well emerges.

We view the universe under the scales made in relativity theory that measure the universes length at 93 billion light years in length. 93 billion light years is also the equation numbers for length of a mass under a gravity field meaning numbers to light speed are engaged by Einstein's light speed constant as well in his GRT and SRT theories but the manner in which he measured our universe.

It's up to us as entrepreneur to believe the universe can have validity to A.E's light constant verities of measuring everything about the universe a side from the universe being trapped in a bubble that is locked in all its behaviors at the velocity of light. Or shut the light out and agree that Einstein set us all up to a host and we should gallantly resume to the better deal.

Albert Einstein used light years instead of billions of miles for the length of the universe because light speed is the greatest length that was known to mankind at the time especially in a universe space field as great as the universe we reside inside. It wasn't until Erwin Hubble whom discovered the universe was actually expanding that meta-parsecs lengths were invented to the equations that a longer length than light years were developed.

The Universe is expanding. It's retaining more and more area causing the galaxies to spread a part at their seams. How long will happen? Forever some believe? There is a limit some say. One of these persons was Albert Einstein. He believed the universe was governed by the velocity of light. That light redeemed everything we need to know about the universe. Though the speed of light is very fast - the universe is more than that. Besides his universe at the end of the ride was a big bang effect - we now know we can travel faster than light due to what's called "A reverse-vacuum expansion constant". That constant was discovered by Erwin Hubble in the year 1938. He is the one physicist that discovered the universe was expanding and measured how fast. Besides Albert Einstein he was the one whom open a new door to the cosmos the one we use today.

But what is a reverse vacuum. It's a time in the duration of the big bang event that the warp age and expansion of a primeval egg exploded and stretched across the cosmos from where it resided outwards to the maximum limit of its existence. As all matter stretched across empty space the fabric of the space it evolved from and like a rubber band stretched out and stretched back from point A back to point B amidst space.

It was this reverse vacuum activity that allowed an empty void to be formed in the middle of that action. When all matter particles and subatomic masses reformed back into gigantic chunks of matter debris at the middle formed a concentrated bubble of pressure that today is what expands the universe. Erwin Hubble explains it as a balloon that is expanding and it seems indefinitely how and why what is written in the great novel about space.

Reverse Vacuum Space Concentrate

A concentrate is a form of substance which has had the majority of its base component (in the case of a liquid: the solvent) removed. Typically, this will be the removal of water from a solution or suspension, such as the removal of water from fruit juice. One benefit of producing a concentrate is that of a reduction in weight and volume for transportation, as the concentrate can be reconstituted at the time of usage by the addition of the solvent.

It will also allow the enlarging of the mass concentrate to increase if it subtle with a continuous flow of cold liquid such as water is igniting its growth or expansion causing the solvent to multiple. The concept of expiration is related but legally distinct in some jurisdictions the removal of water from a solution or suspension, such as the removal of water from a naked fruit of its juice. One benefit of producing a concentrate is that of space pressure reverse vacuum expansion.

Shelf-life depends on the degradation mechanism of the specific product. Most can be influenced by several factors: exposure to light, heat, and moisture, transmission of gases, mechanical stresses, and contamination by things such as micro-organisms the availability of a constant and abundance of supplies. Product quality is often mathematically modeled around a parameter (concentration of a chemical compound, a microbiological index, or moisture content).The universe existing of matter is only 5 % percent. It will keep its best quality for 12 to 49 billion years.

How can a positive energy become dormant meaning that the mass inside the materials object is neither detected or active by some external entity to pulsate a gravity force.? In space, the attraction between masses is bartered by similar mass. Same masses repeal. In space a ship having a positive mass (shell) does the same it repeals existing close masses as stars and planets and repeals their energy.

A ship idle in space than accelerating its mass does not increase due to its un-connection with a secondary energy like on earths as the planets core affects a moving mass on its surface.

In space the surface of space is intermediate to its position on the space grid. The grid surface dopes not retain energy so when the ship starts accelerating how can the hip mass gain energy as relativity states.

Why do objects even having mass sit in space idle from acceleration? It sits because the spaces field the grid its surface is a repulse platform. Our ship sits on it because it repeals masses, energy, anything accelerating and any energy from the ships engines and acceleration mode are dormant from being connected to affect any mass effects of energy mass gain or energy mass matter of the ship.

Mass acted dormant before the big bang event happen and it was the intense and infinitesimal of the bang that cause a cold fusion of energy that generates all matter. In an empty vacuum environment a mass object acts according to its surrounding environment in the same way it takes the planets energy to create a gravitonic connection between the two energy's. In space it is this connection that lacks position when in space.

Changing the Geodesic of Space

We look at space with the idea that space is idle referred to being at rest or not moving.

We look out at space in the same way we do a record player turn-table. The record turn table when the source is turned on starts turning. We look at the turning able in the same way we look at idle space. Our ship on the table sits on the table as it spins and shows to be idle meaning no acceleration. But later when acceleration is enhanced, the ship begins traveling.

These are the circumstances…

What is positive mass

The mass can be positive or negative and is of an object meaning its weight and mass is measured by the force of gravity acting on it. If there is no force acting on the object it has no weight. A mass or object is held together by the force that created it. An object created is considered to be solid, gaseous or light depending on its element. It is the product of what is observed. In universal space zero point density means just that in comparison to the weight and mass of the object or mass and is equal to c^2 which is the speed of light squared, and is held together by the force that binds it.

Mass is therefore, the amount of energy that is held by the single piece of the weighed object but is dependent to the force of the gravity. If the object is in space it weighs zero along with its mass.

Negative mass

Something that causes a negative response or quality

Mass is negative when it retains zero energy it and is not of an energy less than zero. It's something that's activity is preferably the opposite of positive. What this means is that the object is massless, retains zero or no mass, no energy, no weight and is zero and is fluent to an action of repulse and not attraction.

How does all this work?

Positive and negative mass and acceleration occurs when the media it travels through is considered to be nothing or zero and is negative as well and because on the other hand is positive and because of this the object in the media retains the same negative qualities as the field if no gravity exist.

If positive acceleration occurs it means the object is engaged within a field that is positively changed but acts in repulse because the two masses are positive and repel each other. If the acceleration is negative occurs it can only be a combination of positive and negative or negative and positive which occurs an attraction between the two masses and therefore cause dominion interaction. One mass dominates the other dependent its size and mass.

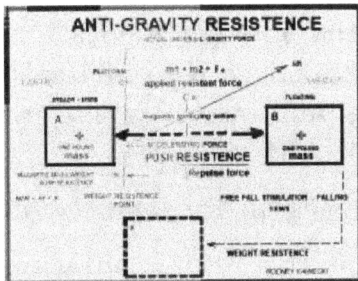

CHAPTER FOUR

Can Gravity be Reversed

What is the Repulsive Gravity Analogy?

Gravity in the quantum vacuum arises from the gravitational repulsion between the positive gravitational charge of matter and the (hypothetical) negative gravitational charge of anti-matter. While matter and antimatter are gravitationally self-attractive, they are mutually repulsive.

What's wrong with space theory definitions:

Both in the Newton theory of gravitation and in the General Theory of Relativity the gravitational force is exclusively attractive one. Why gravity is not repulsive?

There is no way a repulsive force describes that. ... The force is attractive because the phenomenon is one of "accelerating together". There is no way a repulsive force describes that. General Relativity is a better theory, but it has no gravitational force in it. Unlike the Force, with its dark and light sides, gravity has no duality; it only attracts, never repels.

In general relativity, gravity neither pushes nor pulls. ... So in General relativity, gravity is not seen as being a force, instead it is the result objects travelling in the most direct way in a region of curved spacetime.

Yet the evidence shows that a 'gravity force' cannot with strain a positive and negative energy that attracts, the greater the attraction the great the pull the higher the energy content between the masses. The question arises: what keeps planets from crashing together.

A repulse force, the foundation illustrates either poles negative or both poles positive or similar as physics explains for the sources of the gravity force yet contradicts it at the same time. Two a-like poles repel even a subtle free fall for motion in planet atmosphere that gravitate. This is the force of free fall applied without a force. The object just falls. It is the force of acceleration that changes this action. It applies elevation to momentum that changes the rest coordinates of a mass.

Is friction push or pull or drag?

When two surfaces slide over one another the tiny bumps push on each other. Friction causes a force on a surface which is in the opposite direction to its motion.

It is the slowing action of an object with no applied force acting on it - so it slows down to its rest mass state at 9.8 meters a second.

How much pressure is pushing down on us?

At sea level, because of the 60-mile column of atmosphere between you and outer space, there's about 15 pounds per square inch of air pressure pushing down on every part of your body.

What two factors affect gravity?

The force of gravity depends directly upon the masses of the two objects, and inversely on the square of the distance between them. This means that the force of gravity increases with mass, but decreases with increasing distance between objects. The further away the gravitating mass gets between objects the least gravitation. Space gravity is equals to zero.

Does gravity affect weight?

Weight is a measure of how much gravity pulls on a mass or object. On the moon, there is less gravity pulling on objects, so they weigh less.

The most important facts about gravitation is that as a whole the universe creates gravity as the force that allows for motion. On a planet like earths, gravity changes with the energy of the planets mass because it is inside its atmosphere. Without that gravity change objects would fly away slowly like in space adrift. A-drift occurs because gravity in space and the universe also known as 'universal gravity' is directly related to the expanse of the universe. It is that expanse at a constant rate application force that allows or creates motion amongst objects in the universe.

Putting the pieces where they belong

Alan Guth's cosmic inflation theory starts that about 10^{-38} second before the Big Bang which led to our particular universe, a tiny patch of space doubled in size more than 100 times, from sub-atomic size to about 1cm. This "inflation" now ends, the potential energy of the scalar field in this pebble-sized infant universe converts to a hot soup of particles and radiation, and what we call the Big Bang is this hot soup expanding.

In the midst of all this growth in the center of the bang also created a flexible presuming expanse representing an inflating balloon. It is this anomaly that excites and expands the universe.

Dark matter and its affects that generates a push on space are compelled because space lays between a multitude of galaxy's and a expanding probe defined as the cosmic background image. An oval image that physicists believe show what the formation of the universe looks like. A gigantic sphere that was formed at the time of the big bang at time zero has been expanding ever since. If the universe is expanding at a velocity faster than light mankind can never reach its event horizon. A ship traveling at light speed has to match the speed of the expansion and that can only be possible if the ship could travel over 500,000 miles a second and that's without considering the time it takes to reach that velocity.

ATTRACTION

Force of attraction is a force that pulls the body near due to its *attraction*. ... Some of them are magnetic force, electric force, electrostatic force and gravitational force. Gravitational force is very well-identified instance force of attraction as it draws objects towards itself regardless of its distance.

Reversing Newton's Arrows

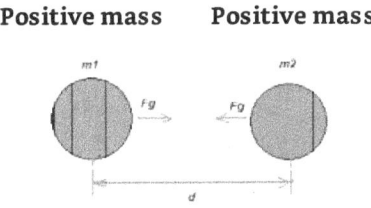

Can Physics be Wrong?

Can Gravity be Reversed?

The force of gravity is an attraction between two objects with mass. In order to have reverse gravity, you would need to have a repulsive effect. So, no. reverse gravity isn't possible.

Why is gravity only an attractive force? Because mass is positive

The force of gravity, FG is proportional to the product of the masses and inversely proportional to the distance, r, squared. ... G is simply a bookkeeping constant that allows us to get the right answer for the force based on any choice of units for mass and distance.

Gravitational force

Gravitational force is an attractive, fundamental force of the universe that is directly proportional to the mass of the two objects and inversely proportional to the distance between them.

The Push Force

In space, in outer space it is not all free fall coordinates that measures how fast a ship can travel at. It is divine intervention flight in zero-point gravity and how the vacuum is. The deeper the area of the vacuum the faster a spacecraft might travel. There is most likely to exist deeper regions of space regions and planetary sections that free fall backed up by propulsion accelerations will push a ship to a most extreme velocity.

Warranting going faster

Infinite Mass: Though many will not agree with me but 'No Such Thing' (except traveling through a planetary gravity force like earth.) Repulse Gravity is a force that acts on a reverse-vacuum activity reluctantly also known as false vacuum state. Mass therefore is being pushed not attracted to create an infinite mass prodigy? Like two positive magnetic – the object mass generates a physical push force between them.

What is the Equivalence of mass, matter and space?

$$m^{\circ}(m^1, m^2) c^2/v^2 = E$$

Seemingly trivial considerations the equivalence of space and mass are the same. Without mass there is no space and just the same without space there is no mass. Mass and space interdependent similar with $E=mc^2$. Thus, the extent of Einstein's universe does not depend on a kind of stable gravitational equilibrium or a kind of balance between the gravitational force of the total mass of the universe a kind of contradicting force that keeps the universe stable, as even wrongly supposed by Einstein. The extent of the universe solely depends on its mass and not on its gravitation caused by this mass. This in turn means, that the energy density of the universe is invariable. The energy density of the universe is solely determined by its mass and by the space defined by the mass due to the principle of equivalency of mass and space meaning that its mass and space are equal in equivalence therefore inner dependent. Therefore, space and mass solely explain the existence of the universe using 'light' as its vehicle.

But there is another side to the coin. Without interaction between the two both space and matter mass does not exist. Einstein explains mass equivalence without the action that 'mass' was formed. He calls matter mass without the assertion to energy and he calls space equivalent to energy without matter.

Interaction takes away the mass equivalence factor meaning the big bang. Mass equivalence exist as a mass equivalence equation without matter. Matter advocates with size as Einstein states space advocates with mass. Meaning that without energy mass does not exist at all.

Einstein's mass equivalence equation is asserted without cause. Cause is what generated mass energy equivalency due to its interaction with space. As such energy does not exist without the big bang.

Interaction raised the value of mass that formed mass after the big bang. Thus it formed the mass equivalence equation.

The idea that space is equivalent with space and space is equivalent to mass is just another way of saying space and mass existed separately - at an earlier time.

The collapse of matter generated the cause for space and mass to interact later thus forming the universe. If space and mass were equivalent in Einstein's view mass and space are equal but space is expanding therefore they are not. Space and mass do not equivalent distance to the size of the universe as relativity equates.

Mass, matter and space are three separate equivalencies' whereas distance can be reviewed with an arrow of time and not absolute as relativity and light portrays the universe.

The collapse of matter generating the mass and space equivalent also generated distance with time as it warps space into a singularity. Without this singularity time in the universe cannot exist in the same way as light is fixed with velocity and time therefore is not equivalent with time.   So what is time as I suggest?

An expanding universe arrowing out the warp-age and generating of its singularity and its expansion without these three suggestions for our universe equivalency the universe does not exist at all.

We could say "space" exist as a catcher's mite, whereby the cosmic egg (matter) was cracked by - caught in the mitt creating the distance for creation" after by the mass was reformed and generated.

Arguments; we review gravitation as the acceleration rate of a mass asserted in free fall. Without an added acceleration so we can observe the trait of what a planets asserts on the free falling mass. When we review the idea that this measurement is in nature of the planet's surface object mobility we can insert an added acceleration and account for what is naturally the planets conservation.

We look at gravitation to be in addition to what the planet conveys naturally to account for how added acceleration can be measured in the analogy. Earlier, physicists assume gravity is an attraction force, like Newton whom measures his laws of motion in detail script but uses a common diagram for his different distinction that are generated by acceleration and in the different categories of his laws.

Cause and effect, the assertion of free fall and acceleration when observed

It is the assertion of 'attraction "that leads physicists especially Albert Einstein to assume where an objects mass becomes a "dragging" affect when the mass or object is accelerated. This is his description of the attraction. But like everybody knows to be factual any two masses or objects of the same positivity stuff made up of particles and elements 'repel'. They don't attract. It is for this reason I have token us to the observation of "repulse".

Repulse affects like observed when two of the same magnetic are placed north pole to north pole or south pole to south pole push away from each other. This encounter is an effect of a natural force acting between the two masses. We know that gravity is a very weak force so weak but it still exists throughout the cosmos. It's what keeps objects and planets a part from each other when they are at rest. The observation of the planets in circular motion and rotations is an alternative route from collision as the planets in the orbitration circle are in actual rest.

There is no external added mechanism that might redirect their pathways other than being at rest.

When we observe the detailed chemistry of two positive electric poles of the two magnetic that act in repel, we discover how it is different from what one observes as attraction. The magnetic are pushed away from each other and it is a natural cause and effect. There exists no attraction between the magnetic.

If we look at this activity in its natural occurrence, we also observe that as two bodies everything conserved in one magnetic has no magnitude or condition applied to with the other magnetic. The intermediate conservation of these magnetic is independent and natural. There exist no entanglement as you would observe with an attraction of two bodies one who's pole is negatively charged and the other positively charged as it would have to be to have an attraction force displayed. This does not occur in a repulse repel interaction as in the attraction force.

Taken all these interactions into account we discover the generated masses and the force implied that separates them is shown to be conclusion and well addressed as slowly the separation that occurs in the repulse natural of conservation between the two masses and the force that separates them illustrate that this natural force of magnetic separation is enforced when an external added acceleration force is applied on one of the masses. In general, are looking at the earth as a whole and a surface object or mass like a rock or a rocket ship that is generating this activity.

As the separation is un-creatively applied by propulsion, the separation is released from the natural occurrence of the example magnetic to full acceleration of the rocket. But when we observe all the facts we observe that the 'attraction "that creates the influence of increasing in the moving object or rocket has been separated by any common denominator that the attraction force encounters in the laws of attraction. Separation is shown to be the true nature of the repulse force in all aspects of the acceleration. Meaning that what happen due to attraction does not happen in the repulse force. The separation and enhancement of acceleration adjoins leaving behind any connection between the interactions of the gravity force behind it therefore procuring separation. .

"Spooky action at a distance"

It's not the first time experiments have demonstrated the spooky phenomenon, known as quantum entanglement. In physics, action at a distance is the concept that an object can be moved, changed, or otherwise affected without being physically touched (as in mechanical contact) by another object. That is, it is the nonlocal interaction of objects that are separated in space. Quantum entanglement is a physical phenomenon which occurs when pairs or groups of particles are generated or interact in ways such that the quantum state of each particle cannot be described independently of the state of the other(s), even when the particles are separated by a large distance—instead, a quantum state

In the case of entangled particles, such a measurement will be on the entangled system as a whole. Given the statistics of these measurements cannot be replicated by models in which each particle has its own state independent of the other, it appears that one particle of an entangled pair "knows" what measurement has been performed on the other, and with what outcome, even though there is no known means for such information to be communicated between the particles, which at the time of measurement may be separated by arbitrarily large distances.

Such phenomena were the subject of a 1935 paper by Albert Einstein, Boris Podolsky, and Nathan Rosen, and several papers by Erwin Schrödinger shortly thereafter, describing what came to be known as the EPR paradox. Einstein and others considered such behavior to be impossible, as it violated the local realist view of causality (Einstein referring to it as "spooky action at a distance") and argued that the accepted formulation of quantum mechanics must therefore be incomplete. Later, however, the counterintuitive predictions of quantum mechanics were verified experimentally in tests where the polarization or spin of entangled particles were measured at separate locations, statistically violating Bell's inequality, demonstrating that the classical conception of "local realism" cannot be correct.

In earlier tests it couldn't be absolutely ruled out that the test result at one point (or which test was being performed) could have been subtly transmitted to the remote point, affecting the outcome at the second location.

However so-called "loophole-free" Bell tests have been performed in which the locations were separated such that communications at the speed of light would have taken longer—in one case 10,000 times longer—than the interval between the measurements. Since faster-than-light signaling is impossible according to the special theory of relativity, any doubts about entanglement due to such a loophole have thereby been quashed.

The idea about spooky at a distance" tries to illustrate the point and behavior of "local realism" the idea that nothing changes over even long distances. However so-called "loophole-free" Bell tests have been performed in which the locations were separated such that communications at the speed of light would have taken longer— which in one case 10,000 times longer—than the interval between the measurements.

The facts rely with the foundation that the smallest particles and in their state are refined to having no possibility of change even experimented at long distances. The facts showed that relevant to "local realism" change occurs in the behavior of smaller particles at great distances. That the average "local realism" is not at all real and by measuring the intervals in distance the energy of a particle will increase differently when separated from its local velocity also establishing FTL is possibility.

This also invited physicists to the realization that change can occur in the natural state of matter and energy in the groups relative to their velocity. Quantum entanglement is a physical phenomenon which occurs when pairs or groups of particles are generated or interact in ways such that the quantum state of each particle cannot be described independently of the state of the other(s), even when the particles are separated by a large distance... particles seem to interact with each other instantaneously, over any distance, breaking the speed of light and thus breaking relativity.

CHAPTER FIVE

Does The Universe Have Mass?

Dirac's Dark Sea

Singular bubble concentrate H^2v

The speed of light squared is a huge number—90,000,000,000 (km/sec)^2—the amount of energy bound up into even the smallest mass is truly mind-boggling.

In miles per hour, light speed is, well, a lot: about 670,616,629 mph. If you could travel at the speed of light, you could go around the Earth 7.5 times in one second.

One of the most famous equations in mathematics comes from special relativity. The equation $E = mc^2$ means "energy equals mass times the speed of light squared." It shows that energy (E) and mass (m) are interchangeable; they are different forms of the same thing.

The Speed of Light is No Longer the Speed Limit

On July 19, 2000 Scientists have apparently broken the universe's speed limit.

For generations, physicists believed there is nothing faster than light moving through a vacuum — a speed of 186,000 miles per second.

But in an experiment in Princeton, N.J., physicists sent a pulse of laser light through cesium vapor so quickly that it left the chamber before it had even finished entering.

The pulse traveled 310 times the distance it would have covered if the chamber had contained a vacuum.

Researchers say it is the most convincing demonstration yet that the speed of light — supposedly an ironclad rule of nature — can be pushed beyond known boundaries, at least under certain laboratory circumstances.

"This effect cannot be used to send information back in time," said Lijun Wang, a researcher with the private NEC Institute. "However, our experiment does show that the generally held misconception that 'nothing can travel faster than the speed of light' is wrong."

Convert units. Noting that $1 \text{ kg} \cdot \text{m}^2/\text{s}^2 = 1$ J, we see the rest mass energy is $E° = 9.00 \times 10^{13}$ J.

The 9.00×10^{13} J rest mass energy for 1.00 g is

Atomic bomb and about 10,000 time's large aircraft

Relativistic ally, at rest we have rest energy $E° = m°c^2$.

Universe pounds and size-mass comparisons

E = approximately 2.246×10^{18} foot pounds of energy to move an average weight person at 99.99% the speed of light.

The Boeing 737-800 airplane is an example of an average sized airplane. It has a maximum takeoff weight of about 80,000 kg (175,000 lbs). This includes the weight of the plane, which is about 41,000 kg (90,000 lbs), and the weight of the fuel which is about 18,000 kg (40,000 lbs).

The 737 Next Generation is made up of the -600, -700, -800, and -900ER models. These models go from 102 ft (31.09 m) to 138 ft (42.06 m) in length

Boeing 737 Airplane (175,000 lbs)

x 2.246×10^{18} foot pounds of energy

=22.4 + 18

=2.24e+18

=3.92e+23

175,000 lbs. Aircraft

Speed of light mass = 391,999,999,999,999,983,222,784 lbs./e

=24.8,770,294,872 pounds

=398.0324717952001 ounces

= 24.8 tons' mass @ 22498.18 kilograms' aircraft mass

As an area twice the size of China

Aircraft Mass

Aircraft size 102 ft (31.09 m) to 138 ft (42.06 m) in length

Mass = area twice the size of China at (42.06 m) in length.

The Universe Expanding Bubble Mass

Mass of universe expanding at 2.7c light speed

=1.0584e +22

2.85768e+22 @ 2.7c

3969000000000 Acre Feet Per Minute

= 5.5870746622583E+22 Gallons [US] Per Month

3969000000000 ac ft/min

= 66150000000 acre feet per second (ac ft/s)

Dirac's Sea of Particle

66150000000 (ac ft/s) e @ Hubble Expansion 42 m/s. Un-concentrating mass expansion into more area…round bubble

Epics of the Big Bang

Inside a whirlpool of an explosion happening in a dense yet flexible state trapped inside like a rubber band stretches outwards to the maximum thread of that energy in a flexible vacuum area reverse engineers back to its original starting point. In such a case was the big bang. The only difference is in this flexible activity and in a three dimensional space having no real place to finish an explosion re-curls back to its original area only to find in the same way as an atomic bomb re-curls the area of inception in space the bang creates a center void absent of anything but emptiness. This emptiness is the direct point to where the unimaginable force returns but not before its infinite heat has melted the dynamic core of the cold dense equilibrium of space.

Like twisting a filled rag condense with water the flexibility of the force of the bang reverse engineered fills this empty area twisting the fabric of space out of its dense water vapor. As a result, it is filled with water. Condense water that affected in the same way as the big bang affected everything of it around reformed the space equilibrium with a bubble recurred by the same density opening the space as the bang not only ob-lit-er-ate all matter and crushed out the cold in the fabric of the equilibrium element with the same maximum dense pressure circumventing a condensed expanding bubble mass invisible in distance and unknown in size so dense it is slowly vaporizing its concentration over time.

From T=0 to today 2021yr.

Zero m/s to 42 m/s decrease of density from zero m/s to 42 m/s. over 13.7 billion years.

In-conclusive to a three-dimensional mass (pi) @ 40.1 billion years

This is the results of that Episode

The vapor pressure of water is the pressure at which water vapor is in thermodynamic equilibrium with its condensed state. At high pressures water would condense. The water vapor pressure is the partial pressure of water vapor in any gas mixture in equilibrium with solid or liquid water.

Pressure will make (something) denser or more concentrated.

Condensation is initiated by the formation of atomic/molecular clusters of that species within its gaseous volume—like rain drop or snow flake formation within clouds.

When there is an increase in pressure like in the big bang, the equilibrium will shift towards the side of the reaction with fewer moles or vapor. When there is a decrease in pressure, the equilibrium will shift towards the side of the reaction with more moles of vapor. In space this would cause a whirlpool effect that diverts the direction of change at the extreme.

Vapor pressure or equilibrium vapor pressure is defined as the pressure exerted by a vapor in thermodynamic equilibrium with its condensed phases (solid or liquid) at a given temperature in a closed system.

The equilibrium constant of a chemical reaction is the value of its reaction quotient at chemical equilibrium, a state approached by a dynamic chemical system after sufficient time has elapsed at which its composition has no measurable tendency towards further change.

The equilibrium constant, K, expresses the relationship between products and reactants of a reaction at equilibrium with respect to a specific unit. This article explains how to write equilibrium constant expressions, and introduces the calculations involved with both the concentration and the partial pressure.

Concentrated or diluted: not really pure water is a pure substance, and there's nothing to remove to concentrate it. It also has very low compressibility, which means that putting pressure one it doesn't increase the concentration much either. You can dilute water out to a very small concentration.

Taking the space element out of the universe's equilibrium core

Dilution is the addition of solvent, which decreases the concentration of the solute in the solution. In such as pressurization and cooling concentration is the removal of solvent (space), which increases the concentration of the solute, in the solution - as the dilution equation.

Like everything that is Ascended in Space

When the volume increases, there are fewer molecules per unit volume. This means the number of collisions per unit of surface area decreases, so the pressure is lower. Effectively, increasing the pressure by decreasing the volume increases the concentration.

The relationship between two solutions with the same amount of moles of solute can be represented by the formula $c^1V^1 = c^2V^2$, where c is concentration and V is volume.

How do you convert concentration to volume?

Divide the mass of the solute by the total volume of the solution. Write out the equation $C = m/V$, where m is the mass of the solute and V is the total volume of the solution. Plug in the values you found for the mass and volume, and divide them to find the concentration of your solution.

The definition of concentration means the amount of ingredients or parts in relation to the other ingredients or parts. An example of concentration is the amount of salt to water in a saltwater solution. .

What is C n V?

"$n = c / v$" "c" is the speed of light in a vacuum, "v" is the speed of light in that substance and "n" is the index of refraction. ... When light moves from one substance to another it changes speed and direction. That change in direction is called refraction.

Concentration is the measure of how much of a given substance is mixed with another substance. For instance, how much space element is mixed with the water core.

Density Equation for these Calculations:

The Density Calculator uses the formula $p=m/V$, or density (p) is equal to mass (m) divided by volume (V). The calculator can use any two of the values to calculate the third. Density is defined as mass per unit volume.

The mass of water is expressed in grams (g) or kilograms (kg), and the volume is measured in liters (L), cubic centimeters (cm3), or milliliters (mL). Density is calculated by the dividing the mass by the volume, so that density is measured as units of mass/volume, often g/mL.

What is the formula for concentration?

The standard formula is $C = m/V$, where C is the concentration, m is the mass of the solute dissolved, and V is the total volume of the solution.

Does increasing volume decrease concentration?

When the volume increases, there are fewer molecules per unit volume. This means the number of collisions per unit of surface area decreases, so the pressure is lower. ... Effectively, increasing the pressure by decreasing the volume increases the concentration.

During remission of pressure space increases and decreases in pressure and its flexibility relevant with the universes cosmic core this activity allows evaporation at the surface, decreasing the volume but increasing the pressure towards the core. It is this activity that increases the cores volume, increasing its mass, decreasing it volume and allows the volume of the bubble to recede. (gets bigger)

What is the formula for area of all shapes?

Area of Plane Shapes

Triangle Area = ½ × b × h b = base h = vertical height   Square Area = a2 a = length of side

Rectangle Area = w × h w = width h = height Parallelogram Area = b × h b = base h = vertical height

Write down the formula for finding the circumference of a circle using the diameter. According to relativity the diameter is 93,000 billion light years.

The formula is simply this: C = πd. In this equation, "C" represents the circumference of the circle, and "d" represents its diameter. That is to say, you can find the circumference of a circle just by multiplying the diameter by pi.

answer: 279,000 billion light years round oval cosmic background imagery universe

What is a formula of a circle?

The center-radius form of the circle equation is in the format (x – h)2 + (y – k)2 = r2, with the center being at the point (h, k) and the radius being "r". This form of the equation

π (Pi) times the Radius squared: A = π r2. or, when you know the Diameter: A = (π/4) × D2. or, when you know the Circumference: A = C2 / 4π

Any prism volume is V = BH where B is area of base and H is height of prism, so find area of the base by B = 1/2 h(b1+b2), then multiply by the height of the prism.

Why do we use PI in Circle area?

Pi r squared

In basic mathematics, pi is used to find the area and circumference of a circle. Pi is used to find area by multiplying the radius squared times *pi*. Because circles are naturally occurring in nature, and are often used in other mathematical equations, pi is all around us and is constantly being used.

What is the formula for speed?

To solve for speed or rate use the formula for speed, s = d/t which means speed equals distance divided by time. To solve for time, use the formula for time, t = d/s which means time equals distance divided by speed.

What is the total surface area of a sphere?

The curved surface area of a sphere is 4(pi)r^2 or (pi)d^2.

To find the surface area of a sphere, use the equation 4πr2, where r stands for the radius, which you will multiply by it to square it then, multiply the squared radius by 4. For example, if the radius is 5, it would be 25 times 4, which equals 100.

What is the formula of hollow sphere?

To find the surface of the sphere we must coordinate from the volume of the sphere,

To determine the volume of a sphere, we use the formula 4/3πr^3. So, to find the volume of a hollow sphere, we must subtract the volume of the hollow region from the volume of the overall sphere. Let's call the radius of the overall sphere r1 and the radius of the hollowed region r2. Thus, we get 4/3π*r1^3 - 4/3π*r2^3.

Surface area of a sphere: A = $4\pi r^2$, where r stands for the radius of the sphere. Surface area of a cube: A = $6a^2$, where 'a' is the side length. Surface area of a cylinder: A = $2\pi r^2 + 2\pi rh$, where r is the radius and h is the height of the cylinder.

The mass and volume of the sphere, it's size-mass to intensify; make denser, stronger, or purer, especially by the removal or reduction of liquid.

Why would the expansion pod wobble unless to cause flexibility against the vacuum equilibrium and release tension to allow evaporating that allows the core to enlarge or expand when at the same time creating a friction that allows the core to manipulate the vacuum to conserve concentrated liquid-space and energy aberrations.

The Universe's Expansion Core

The Primordial Galaxy

Trying to measure the age of the universe by measuring the age of the oldest star is like forgetting that the stars and planets were formed in the galaxies. That the galaxy's need not be explored for data that constitutes valuable information that goes back further than a single planet or star they reformed and overtime. Radiation that warms the chicken egg is the same radiation that created birth of the galaxy's that gave birth to the planets that gave birth to the stars and before their time and age.

The primordial expanding egg or bubble made at the time of the bang could be a vaporizing frozen paramount. The expansion can be detailed as hard frozen watered solution mixed with equilibrium element used as a glue because of the temperature outside in idle space slowly evaporates into water than vaporizes adding to the space area but is expanding in mass because it is changing form from infinitely dense ice to water rippling its surface as it slowly vaporizes into nothingness. And like all liquids in space vacuum it stays round shaped stabilized due to the surrounding pressure outside the bubble when everything rushed back after the extreme force reached maximum potential and stretched back again under the influence of a dense flexible equilibrium. The erecting area of the bang filled with the shrinking space fabric having change rapidly under the pressure allowing the area to be filled with water.

The Center of the Mass

Space is a lot of things: cold, dark, and empty come to mind right away. And those things become pretty evident quickly, as soon as you leave the Earth.

The temperature of space is, at its coldest, just the temperature of the leftover glow from the Big Bang. This radiation, known as the Cosmic Microwave Background, bathes the entire Universe in a temperature of only 2.7 Kelvin. If you can adequately shield yourself from the Sun, the Earth, and all other sources of heat in our local neighborhood, that's how cold space is!

That temperature is less than 3 degrees above absolute zero, or -455 degrees Fahrenheit, so it's vital when we send humans to space to keep both appropriate temperatures and pressures for them to survive. In a sealed environment like aboard the international space station, water behaves pretty similarly to how it does on Earth, but rapidly and with the remarkable exception of gravity.

But if you were to put some liquid water out into deep space with its freezing temperatures, would it freeze? Remember that there's also literally zero pressure in space. So, what happens? Who wins? Does the water freeze from the low temperatures, or boil from the lack of pressure? The burn off is very slow as all the suns are breaking into divided pieces.

Oddly enough, the answer is the first one, and then the other!

It turns out that having a pressure vacuum result in an incredibly rapid transition, causing the water to boil almost instantly. The (formerly) liquid water has no choice but to enter the gaseous phase, while it will take quite some time for its temperature to drop significantly enough to transition into the solid phase. In other words, the effect of boiling is much, much faster than the effect of freezing under these conditions.

For the universe to contain an infinite account of strained water out of the big bang it would rapidly noting that we are reviewing a vacuum time limitation, and change it to ice that slowly vaporizes at the Hubble constant.

But the story doesn't end there. Once the water had boiled, we now have some isolated water molecules in a gaseous state, but a very, very cold environment! These tiny water vapor droplets now immediately freeze (or, technically, DE sublimate), and become ice crystals.

We've actually observed this before. According to astronaut observations, where they've observed their urine getting expelled from the spacecraft:

When the astronauts take a leak while on a mission and expel the result into space, it boils violently. The vapor then passes immediately into the solid state (a process known as DE sublimation), and you end up with a cloud of very fine crystals.

You can define a center of mass. If an object is finite, the center of mass is just the point that, on average, has an equal amount of mass surrounding it and in all directions. The situation gets more complicated for an infinite object. If an object is infinite and uniform, you simply cannot define a center of mass because all points are identical. On the other hand, if an object is infinite but not uniform (for instance it has a single knot of high density at one point), you can define the center of mass of the entire object as the center of mass of the non-uniformity.

These observations are the foundation for the concept that a Big Bang started the universe. Because the universe is expanding, if you run time backwards, there had to be a time when the universe was all compacted to one point. Since the universe is expanding, you would think there is a center of expansion. But observations have revealed this not to be the case. The universe is expanding equally in all directions. All points in space are getting uniformly distant from all other points at the same time. This may be hard to visualize, but the key concept is that objects in the universe aren't really flying away from each other on the universal scale. Instead, the objects are relativity fixed in space, and space itself is expanding.

The Big Bang is the center of the universe. But because space itself was created by the Big Bang, the location of the Big Bang was everywhere in the universe and not at a single point. The major

THE COSMIC CORE

aftereffect of the Big Bang was a flash of light known as the Cosmic Background Radiation. If the Big Bang happened at one location in space, we would only see this flash of light coming from one spot in the sky (we can see a flash that happened so long ago because light takes time to travel through space and the universal scale is so big). Instead, we see the flash as coming equally from all points in space. Furthermore, once the motion of the earth is accounted for, the flash of light is equally strong in all directions.

The universe simply has no center. The universe is infinite and non-rotating. Averaged over the universal scale, the universe is uniform.

It is said that everything spins in the Universe – from galaxies to stars to planets. ... This spinning of galaxies continues even after their formation. Our Milky Way galaxy is one of these spinning structures and its entire disc of stars, gas and dust is rotating at around 168 miles per second. All galaxies are in orbitration but what makes them spin?

Yes, gravity can form waves. Gravitational waves are ripples in space-time that travel through the universe. If you think of gravity as force acting at a distance, it is difficult to visualize how gravitational waves could form.

Michio Kaku defined gravity as ripples in a sea. Could he be right? That incidental the primordial egg is made of water solution possibly everything that comes with a dying universe.

News confirms it

The background radiation measured today is at 2.7k the Universe cooled to this density will slowly change. Radiation is a light weak mass that fills the dormant area of the whole universe as it stretches across the cosmos. For weak filler, we can compare it with the core speed limit or density that the universe's "G" gravity allows for acceleration or a speed limit we cannot travel faster than the universe allows us. That density is embedded with the vacuum density that the 2.7k is measured at. We can assume that since space is empty relative to any dominion interaction allowed by the universe, which since the universe is in expanse it is an open system but accounts to maintain a cosmic background radiation density we can read. Along with this hypothesis we can verily assume that if its equal with the universe's expansion velocity that tells us how fast the universe is expanding that these two numbers are be assumed common knowledge in the hitherto. Since the universe is expanding at 2.7c, and the cosmic background radiation measures 2,7k that the room that exists as a vacuum is contained even though it's an open system. It was good when the universe was younger it was smaller and denser. The universe will get cooler and does at the same velocity as Hubble's balloon or Stephen Hawking' bubble expands. It' is at the center of the universe this balloon exists and is slowly vaporizing as it keeps the universe at a special temperature and orbital continuum. It's the only thing that stays the same throughout the universe. For acceleration throughout the universe 2.7 k temperature changes but slowly with the same content the universe expands. The cosmic background radiation seems to be the hierarchy boiling point for all planetary matter throughout the cosmos. But what does all this mean when we talk about hierarchy radiation as a universal constant and why.

This means a ship made of mass same as everything else will boil over at a specific point in equal to the background radiation. Reluctantly in 1905 Einstein review this measured at the speed of light. History doesn't change with the new findings. For advance physics to become changed in the history books is a valued reward and achievement.

Reluctantly the measurement predictions are wrong when reviewing the facts about the cosmic background heat. The universe is expanding no more than what it was at the time of the big bang thirteen billion years ago. and at the speed of light, to that comment light is a fixed number for light to be a universal constant and of course to stop matter but matter seems to have a universal boiling point.

Cosmic background radiation is the way that they measure the age of the universe. The universe is 14 billion years old according to the equation. And there's no way to get around these figures. But again we emphasize to the fact that it is that the speed of light constant is the reason that the universe reluctant to all the chemistry that plays a part in our universe and it's on changing expansion that Einstein's universal speed limit is invalid. But the universe is expanding and so this number changes. We are not stuck on 'c'. The universe gets cooler due vapor rising from universal bubble or void. So we are set whether or not the universe will get colder or get hotter or to stay the same as it is. What cosmic background radiation will not do is change the age of the universe or so it seems.

We look at the beginning of our universe way back to the time of the Big Bang really and reluctantly we look at the energy involved in that Big Bang. Physics audience says the universe is two-dimensional flat meaning it has no real curvature but update information to Hubble laws of the universe is round like a balloon. With a round universe there's more matter meaning more mass but does that change the readings of the 2.7k radiation. No it does not.

Here are the numbers, 450,000 miles per second where every second the universe mass is changing. So the figures being correct then our speed limit changes to two points in 2.7c. And that is the new speed limit for everything.

Where the universe boiling point for all matter is 2.7k the radiant atmosphere for the universe it evolves to the 2.7c expansion. Pressure of the universe is an equal comparable with its counter point boiling point where these two figures may show better the circumstances of a universal gravity speed limit to be correct.

Albert Einstein's channel to the cosmic background radiation as a standard he measured it to be relative to the speed of light but over time we have learned that the universe is forever changing. Even though the speed of light doesn't change if the cosmic speed limit is greater than the speed of light it's just a question of how much greater. The quanta physics theory shows that it is 2.7c not just 'c' but what does that a constant rolling over time mean.

One thing for sure the universe is not static and it's not constant where the speed of light in relativity it is and the variable matter plays as its own cosmic background Albert Einstein's equation for mass in a universe which is forever changing to the values that no longer apply. It is realizing that radiation is a boiling point relative 2.7c and not just the speed of light or infinite mass equations that by reconcile to this new intervention tells us where the universe evolution will take us all as a species.

We should look at the parable that space by itself has no real mass. That it coordinates fects relative to any activity that corresponded in its presents. The universe is expanding but by what means is it. There exist chemicals like cosmic background radiation that play a special role in how the universe was formed but still these chemistries are separate deities to debris that exist and not relative to space field alone.

Where the relativity infinite mass energy theory illustrates a limit to the input of energy in a moving vessel at the speed of light, if scientific theory proves faster than light space travel is a fact in future space exploration like 'The Quanta Physics Theory' shows than the infinite energy c/v technology is wrong. Whereas cosmic background radiation abides by the same rules as infinite mass theory and energy theory they too should follow behind.

Since this faster than light technology propaganda stems from how much gravitation in space acts on moving objects where the universal speed limit is shown to be different or incorrect and a new speed limit relevant to universal gravity can apply instead for space and how it applies to acceleration can be view as the remedy and the height of acceleration occurs relative to the background radiation paradox the strength of the paradox changes so does the application of the theories involved.

Why Time Travel in Relativity Theory Fails

In the beginning time travel was a gateway to visitors from outer space and tales about how it would be if time travel were possible. But later in 1904 a man name Albert Einstein invented a theory called Special Relativity. Part of that theory supposed that if a person could travel at close to the speed of light he would actually travel into the past. His theory was physically based on the idea that at the velocity of light was the fastest anyone or anything could travel on any type of journey travelled through time. His facts slowly came to be believed to true when he illustrated how the universe had a specific speed limit. This speed limit was the fastest anything can travel but with it also invented the idea that if mankind could bend space, slow down planetary orbits or built a white wormhole that these were portals that a ship could travel through and adventure time in the past or in the future. If a ship could travel at light speed he or she could travel directly into the future because to travel into the past the ships driver would have to travel faster than light. All these analogies lead people to believe time travel was possible.

The other idea is of Einstein's were that if mankind could build a wormhole, bend and curve space they would travel faster than light than be able to travel into the past. But all these speculations were based on an idea that light speed is the fastest anyone or anything can travel. His theories also were able to penetrate the cosmic background radiation theories that energy as well as any material mass was also limited to light speed. This idea was also based on the universal speed limit in his theories.

It wasn't until about 1942 that a physicist Erwin Hubble discovered venturing through a telescope at star imagery showed him the universe was expanding. Researching photographs of galaxies over periods intervals of days and even weeks showed the galaxies were spreading a part from each other. This meant the stars in the universe were not fixed as Einstein pursued it. What this suggested was that the universe was expanding getting bigger and bigger over time which also meant Einstein's theory about a universal speed limit was wrong. It also showed the universe was not static as well as that universe expansion meant that according to relativity if the universe was static and analogies at the speed of light that traveling faster than light was possible. That according to the cosmic background radiation analogy the universe did not venture a speed limit at all, that faster velocities were possible solely based on the facts that the universe was expanding and that that expansion was constant.

So what did this mean for time travel? It meant that things did travel faster than light as suggested because the universe was not fixed at light speed until after it cooled and things such as a mass were not subject to the Einstein speed limit meaning that the universal speed limit does not interfere with space travel because there was no speed limit. That time travel relied on a speed limit Einstein invested with but that that speed limit wasn't a reality. It means Einstein was wrong. Thus his time travel theory was wrong.

Because all Einstein's adventures relied on his universal speed limit his theories do not pan out as true but in all his ventures did accomplish the idea about a mathematical sum in $E=mc^2$, but even so the numbers can be wrong as well if we cannot establish a speed limit and a speed limit relies on a new theory.

Reading this book surely hasn't left behind you a spirit un-imagined. In this book exist a new theory based on the analogies about a different space equilibrium that offers space exploration a new Travelocity. A new theory that opens the books to faster than light space travel theory that fits all the proposed media that is necessary to reform advance physics.

There are a few fundamental facts about the Universe its origin, its history, and what it is today that are awfully hard to wrap your head around. One of them is the Big Bang, or the idea that the Universe began a certain time ago: 13.8 billion years ago to be precise. That's the first moment we can describe the Universe as we know it to be today: full of matter and radiation, and the ingredients that would eventually grow into stars, galaxies, planets and human beings. So how far away can we see? You might think, in a Universe limited by the speed of light that would be 13.8 billion years: the age of the Universe multiplied by the speed of light. But 13.8 years is far too small to be the right answer. In actuality, we can see for 46 billion light years in all directions, for a total diameter of 93 billion light years.

The longer we wait, the farther we can see, as light travels in a straight line at the speed of light. So after 13.8 billion years, you'd expect to be able to see back almost 13.8 billion light years, subtracting only how long it took stars and galaxies to form after the Big Bang.

As the Universe expands, the fabric of space stretches, and those individual light waves in that space see their wavelengths stretch as well. And why because the universe is rounded by a center of mass core that is making it expand. Put that all together, the distance we can see in the Universe, from one distant end to the other, is 93 billion light years across but this in a flat 2 dimensional universes. If the universe is round and relativities basis for the speed of light are wrong in a three dimensional universe the universes mass is three time greater not less than as Einstein projects his two-dimensional universes mass. Its mass is three times greater.

So 92 billion light years might seem like a large number for a 13.8-billion-year old Universe, but it's the right number for the Universe we have today, full of matter, radiation, dark energy, and fused matter all obeying the laws of General Relativity. The fact that space itself is expanding, and that new space is constantly getting created in filling in between the bound galaxies, groups and clusters in the cosmos, is explaining how the Universe got to be as big as it is to our eyes.

But for a three dimensional expanding universe with three time the greater mass that the universal speed limit re buys light velocity into a mass for the universe at 92 billion light years across, it illustrates an expanding universe retains the greater value in its own mass equations at three time greater the mass showing an older universe than 13.8 billion year age but having a greater physical mass with the same subtle cosmic background radiation background, its more simpler to agree that the universe was created from a greater bang than the velocity of light portrays to us today in modern physics. Even though light does have its fixed quality's absorber in a vacuum can exceed the local speed limit intensifying the mass. The greater excessive speed limit reformed by a 279 billion light year circumference sound the more logical and an older universe's age at 41 Billion Years Old.

The reciprocal:

For every action there is an opposite reaction in the opposite directioN

{Does not occur in both Newton's dialogue or Einstein's}

Pos. magnetic + Pos. magnetic = is a repulse reaction.

The Repulse Force

Earth mass + Moving Surface Mass

P + P = 9.8 m/s Repulse

Pos. / pos. = repulse

Einstein

Pos, + Neg. = 9.8 m/s attraction

Newton

Pos. + Neg. = 9.8 m/s attraction

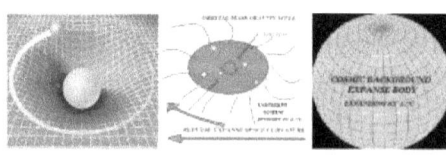

Pi to a few digits

3.14159265358979323846264338327950288419716939937510582097494459230781
6406286208998628034825342117067982148086513282306647093844609550582231725359408128481117450284102701938521105559644622949489549303819644288109756659334461284756482337867831652712019091456485669234603486104543266482133936072602491412737245870066063155881748815209209628292540917153643678925903600113305305488204665213841469519415116109...

By the rule of the thumb, an object isn't falling when it is in 'freefall' because it is gathered up and lifted by the gravity force. In space that same lift that gathers an object in a gravity force is the same lift that gathers that same object in space and keeps it there.

It is the weight of the object that doesn't change because it is 'pushed' or accelerated by a rocket force. Whereby, an object in freefall has no velocity acting on it when in freefall, it retains no weight where in space it is the same. When the acceleration is applied or the rocket it doesn't change the weight of the object or ship because it has already been measured by space. So, it is just accelerated.

If the universal gravity force is a repulse force, the magic push between the masses relative to freefall is slowly diminished because of the force of gravity that when the object is at rest finishes.

Space does not have a dominion sun to advocate its actions. It has nothing but empty space in abundance and distances. Space has defined its presents and its weight (mass) as zero in respect to itself. And to what that tells us about acceleration and velocities is beyind common knowledge.

CHAPTER SIX

Can We Reverse Gravity

A ship travels through idle space. It has the capacity to travel at close to the speed of light. Because of space expansion due to our expanding universe the ship travels 372,000 m/s. The Reverse Vacuum Effect caused by the flow of the universe's expansion allows zero mass ($m°$) receding vessels traveling through space be pushed away furthermore rather than be attracted by close planetary gravitation and so let's us travel through an idle space field at 99.999 percent capacity. This also includes faster than light velocities with no speed limitations because the reverse vacuum field gravity uses the equation $E=m°G^2$, instead of the latter $E=m°c^2$.

The reason is rather that gravitation in space is a reverse-vacuum natural dark matter push force, and rather hopelessly a physical opposite of the planetary attraction force.

With no exterior forces acting on our vessel except for the universes space expansion volume force it we are able to travel the full acceleration value of 99.999 percent velocity having nothing there to slow us down. Venturing into relativities infinite mass theories, we resume the value of the uncritical assumption of 'nothing can travel faster than light" due to the attraction field forces to the indifference principle with the expansion value that the ship that travels light speed due to the conditions of reverse vacuum response now travels twice that velocity based on the standard nature. What attraction theory does to slow things down the reverse vacuum response accelerates thing up faster.

As the capacity of acceleration rises due to the advancement of technologies our universe will continue to expand getting larger and larger over time. The idea that we will ever be caught up with god's creation is overwhelming. Today big "G' allows us to travel 450,000 miles in a single second of time but in the future this number will change. The idea that mankind reaches all the criteria in its engineering propulsion engines that will exceed this number relies on the idea that a reverse push vacuum expansion force allows an expansion of primeval accelerations can be measured with the foundation that the universe space field comes with its own advance acceleration condition that when or if reached will take more propulsive engineering advancement technology to oversee the greater velocity.

The Alcubierre Warp Drive

In 1994, Alcubierre proposed a method for changing the geometry of space by creating a wave that would cause the fabric of space ahead of a spacecraft to contract and the space behind it to expand. The ship would then ride this wave inside a region of flat space, known as a warp bubble, and would not move within this bubble but instead be carried along as the region itself moves due to the actions of the drive. It was thought to use too much negative energy until Harold Sonny White said that the amount of energy required could be reduced if the warp bubble were changed into a warp ring.

The Alcubierre drive, Alcubierre warp drive, or Alcubierre metric is a speculative idea based on a solution of Einstein's field equations in general relativity as proposed by Mexican theoretical physicist Miguel Alcubierre, by which a spacecraft could achieve apparent faster-than-light travel if a configurable energy-density field lower than that of vacuum (that is, negative mass) could be created.

Rather than exceeding the speed of light within a local reference frame, a spacecraft would traverse distances by contracting space in front of it and expanding space behind it, resulting in effective faster-than-light travel. Objects cannot accelerate to the speed of light within normal space-time; instead, the Alcubierre drive shifts space around an object so that the object would arrive at its destination faster than light would in normal space without breaking any physical laws.

The theoretical physicist Miguel Alcubierre was born in Mexico City, where he lived until 1990 when he traveled to Cardiff in the UK to enter graduate school at the University of Wales. He received his PhD from that institution in 1993 for research in numerical general relativity, solving Einstein's gravitational equations with fast computers. He continues to work in this field, devising numerical techniques for describing the physics of orbiting black holes that spin down to collision.

Alcubierre published a remarkable paper which grew from his work in general relativity, the current "standard model" for space-time and gravitation. His paper describes a very unusual solution to Einstein's equations of general relativity, described in the title as a "warp drive", and in the abstract as "a modification of space time in a way that allows a space ship to travel at an arbitrarily large speed". In this Alternate View column, I want to explore Alcubierre work and its implications.

There is also a second faster than light (FTL) prohibition supplied by special relativity. Suppose a device like the "ansible" of LeGuin and Card were discovered that permitted faster-than-light or instantaneous communication. Special relativity is based in the treatment of all reference frames (i.e., coordinate system moving at some constant velocity) with perfect even-handedness and democracy. Therefore, FTL communication is implicitly ruled out by special relativity because it could be used to perform "simultaneity tests" of the readings of separated clocks which would reveal the preferred or "true" reference frame of the universe. The existence of such a preferred frame is in conflict with special relativity.

General relativity treats special relativity as a restricted subtheory that applies locally to any region of space sufficiently small that its curvature can be neglected. General relativity does not forbid faster-than-light travel or communication, but it does require that the local restrictions of special relativity must apply. In other words, light speed is the local speed limit, but the broader considerations of general relativity may provide an end-run way of circumventing this local statute. One example of this is a wormhole [see my AV columns in Analog, June-1989 and May-1990] connecting two widely separated locations in space, say five light-years apart. An object might take a few minutes to move with at low speed through the neck of a wormhole, observing the local speed-limit laws all the way. However, by transiting the wormhole the object has traveled five light years in a few minutes, producing an effective speed of a million times the velocity of light.

Another example of FTL in general relativity is the expansion of the universe itself. As the universe expands, new space is being created between any two separated objects. The objects may be at rest with respect to their local environment and with respect to the cosmic microwave background, but the distance between them may grow at a rate greater than the velocity of light. According to the standard model of cosmology, parts of the universe are receding from us at FTL speeds, and therefore are completely isolated from us. As the rate of expansion of the universe diminishes due to the pull of gravity, remote parts of the universe that have been out of light-speed contact with us since the Big Bang are coming over the light speed horizon and becoming newly visible to our region of the universe.

Alcubierre has proposed a way of beating the FTL speed limit that is somewhat like the expansion of the universe, but on a more local scale. He has developed a "metric" for general relativity, a mathematical representation of the curvature of space that describes a region of flat space surrounded by a "warp" that propels it forward at any arbitrary velocity, including FTL speeds. Alcubierre warp is constructed of hyperbolic tangent functions which create a very peculiar distortion of space at the edges of the flat-space volume. In effect, new space is rapidly being created (like an expanding universe) at the back side of the moving volume, and existing space is being annihilated (like a universe collapsing to a Big Crunch) at the front side of the moving volume. Thus, a space ship within the volume of the Alcubierre warp (and the volume itself) would be pushed forward by the expansion of space at its rear and the contraction of space in front. Here's a figure from Alcubierre paper showing the curvature of space in the region of the travelling warp.

For those familiar with usual rules of special relativity, with its Lorentz contraction, mass increase, and time dilation, the Alcubierre warp metric has some rather peculiar aspects. Since a ship at the center of the moving volume of the metric is at rest with respect to locally flat space, there are no relativistic mass increase or time dilation effects. The on-board spaceship clock runs at the same speed as the clock of an external observer, and that observer will detect no increase in the mass of the moving ship, even when it travels at FTL speeds. Moreover, Alcubierre has shown that even when the ship is accelerating; it travels on a free-fall geodesic. In other words, a ship using the warp to accelerate and decelerate is always in free fall, and the crew would experience no acceleration gee-forces. Enormous tidal forces would be present near the edges of the flat-space volume because of the large space curvature there, but by suitable specification of the metric, these would be made very small within the volume occupied by the ship.

All of this, for those of us who would like to go to the stars without the annoying limitations imposed by special relativity, appears to be too good to be true. "What's the catch?" we ask. As it turns out, there are two "catches" in the Alcubierre warp drive scheme. The first is that, while his warp metric is a valid solution of Einstein's equations of general relativity, we have no idea how to produce such a distortion of space-time. Its implementation would require the imposition of radical curvature on extended regions of space. Within our present state of knowledge, the only way of producing curved space is by using mass, and the masses we have available for works of engineering lead to negligible space curvature. Moreover, even if we could do engineering with mini black holes (which have lots of curved space near their surfaces) it is not clear how an Alcubierre warp could be produced.

The Time It Takes to Travel There"

Is time travel possible? Some think it is - but really is it? The fourth dimension 'demeanor' measures by the threads of time as a dimension of itself. But space-time is based on the reality of the universes motion which is real and sits independent of it being it is the universe itself. To say time is an actual dimension that interchanges within itself past, present and future actually means time is absolute and nothing in it can change. So time travel is impossible. (It either is(present) was(past) or never will be (future) because these aspects about reality and the universe are absolute. They are either present or they are not. Time travel would then have to come from the future earth and it doesn't exist at all except theoretically.

The question at hand is asked "What is Gravity"? The following question arrives "How does it work? To some persons of importance on this planet magistrate like Queen Elizabeth was very puzzles about trying to answer this question. In 1667 the queen assigned her imperialist Isaac Newton to research out answers to the mystery about gravitation. She asks Isaac "why do the apple fall to the ground'? The question should have read "Why do the apple fall from the tree" as a result for his task Isaac Newton discovered a mathematical sum that seemed consistent for answering the question about earth gravity (a truthful and correct answer should have read 'The apple falls because the apple has reached maturity raised from the planets natural elements – my queen).

Isaac measured the velocity and action of free falling objects. He discovered that no matter what the weight or size of an object mass – all masses seemed to free fall at 9.8 meters per second no matter what. The only indifference to this consistency of free fall he called 'attraction" or objects falling freely towards the earth's surface was the inconsistency of added force acting

No matter what – inconsistency was added by a push or acceleration overwhelming direction and pushing force behind an object in motion and changed the consistency of the objects free fall adding an additional pushing force to a freefalling object falling towards the planet's surface actually created a maneuvering passiveness to form by the freedom of the objects free-fall mobility. An object aviates by free fall passing through the earth's atmosphere measured only by the earth's atmospheric gravity dome or gravity ring. The part of the planet's atmosphere or bubble concealed by the planets momentum impressed due to the planetary orbital cycle mobilized within the fabric space grid plate.

As the earth rotates it is pressed physically against the fabric space grid. Because of the universe orbitration energy all galaxies as well as planets and star spheres embed themselves pressed and engraved against the space fabric itself. Because of the existence of matter formed by the big bang event that happens in all universe birth events throughout empty space the dark fabric by which the planetary matter laying relevant to this first beginning event is what generally mobilized a material space grid. Everything that exists in the universe was made in and from it thereby lies within its grid age.

While the universe was set in motion by the first big bang even impression of the first universe what followed was a good weight age of galactic and solar star system little bangs. All these following little big bangs all resulted by the same means as the first big bang in nature. The beginning big bang event started the grid motion in the dark element started by the inconsistency of what we might try and measure as immeasurability. While the massive first universal event mobilized an orbitrating cycle between matter and the dark element of space informing it into a grid smaller and shorter little bangs reformed the first event creating new and deeper grid plate age inside the newly formed universe. Inside these smaller galactic bangs even smaller and denser tiny bangs proceeded. Over the immeasurable time existence preceding us after the big bang event ordainments like Christmas lights sometimes proceed us outside are planets embedded space grid opening and allow us to observe what are called supernova events. Planetary stars and or possibly a system of stars or planets appearing from nowhere amongst the planetary observation we have already mapped of our solar system.

As you can see – the space grid gravitation element along with planetary gravity like here on earth measure and physically appear quite indifferent and they are.

Isaac Newton researched and deciphered the thread of object free fall passing through the earth's atmosphere at 9.8 meters per second. He also maintained that objects in free fall towards the earth's surface were an attraction. He led on to say that the attraction was formed by and objects mass energy that is weighed inside the molecular realm of all matter. He deciphered that in the same way a magnetic attracts other entities objects falling towards the earth's surface are attracted in the same way. It was later that Albert Einstein pursued the theory of gravity and assumed the idea that this same energy mass in dominate objects would built infinite mass in acceleration. That a moving object traveling at close to the speed of light could maneuver no faster due to the lack of energy. Relative to special and general relativity research – the speed of light was constant. Nothing could maneuver faster.

Travel coordinates for the 14 light years to get to the Nearest Star

That's 14 years at light speed

Can we get there faster ?

A Quanta Physics Theory

The figures are shown here:

93 billion light years (flat universe) astronomy dictionary Albert Einstein GRT

Round-about three-dimensional mass – a multiple of pi.

279,000,000,000 light years (^{2}c) pi

3,000,000 light-years (minus a zero point for gravity) to maintain a substantial rate of velocity)

10,000 x 45 miles a sec (46.2 plus or minus 1.3 miles) = 450,000 miles a sec

Without using light frame measurement intervals"

(450,000 m/s) divided by L.S. = 372,000 m/s plus 78,000 m/s equals Speed without Limit

Equals: 44.7 miles per second every 3 million light-years

{450,000 m/s @ 99.999999 = 372,000 m/s} $E=m^°g^2$ (2.7□)

Distance to Neatest Star

14 light years to nearest star = 8,189,475,840,000 miles away

One light year = 31,449,600 seconds = 584,962,560,000 miles

$E=mg^2$ @ 372,000 m/s

Today's modern calculation to the nearest star using relativity theory

Actual 7 year journey to nearest star

Days, months and years to neatest star

8,189,475,840,000 miles to nearest star

18,198,835 seconds @ g^2

303,314 minutes @ g^2

5055.2 hours @ g^2

210.6 days (to neatest star)

Using The Quanta Physics Theory in retrospect [a difference in 'G' matter's {2 x L.S. plus .7c}

= 7 months (to neatest star) @ g^2, defines as how fast the cosmic core is expanding a second.

$$p^{e2} = m^o g(2.7c)$$

To travel faster than that means having the propulsion/ energetic capacity to do it

Warp Drives

The ability to increase velocity in major units of warp drive at 186,000 miles per second per unit for each warp drive

You do not gain mass. {light mass (zero object + the mass @ (light-mass)=zero

No polarity exist in object masses

Matter does not conform to a type of voltage deterrent its ridicule is constant,

$c^1 + c^1 = c^2 (2 \, Lm)$?

c^2 = c squared. {Generating mass - like moving electricity = zero.

In preference to 'light' ;e;, light, 'm',, *all* = zero in vacuum

Reference : light quantity of any mass at *light speed*

{Einstein infinitesimal relativistic equation backwards}

The universe has an expanding gravity force with a velocity equal to 372,000 m/s plus 78,000 m/s or to be more exact in numbers: 450,000 miles per second is the accelerating speed of gravity a carrier force that continues to grow greater over time. All this in accord with the common standard theory about the universe as we know it there are two different origins about the position of the universe. The first is its matter oscillates inside an expanding bubble formation. The second is that matter oscillates outside the inflation of growing pressure. We will be deciphering the universe oscillating outside the bubble whereas the standard theories about the universe all maintain their sufficient is operating inside the bubble formation. This indifference between the two universe theories of sufficiency and universe structures science have only researched using the inside bubble theory over the decades whereas Quanta Physics is the first theoretical science that describes the universe outside the inflating universe bubble pressure and expansion observing all the common laws of physics which render The Quanta Physics' Theory a new theory of advance physics.

We look at two parts about the universe little has been written about. The first is the earlier fate of its origin and the second where the speed of light stands between both of them. We look at the earliest avenue about the universe's origin and discover that in part of its collapse as a mature cosum the universe underwent a closed space barrier explosion. As part of its collapse the universe expanded in for the most part of itself to a degree it expanded beyond the density the universe is measured by today. As a closed eruption there existed no escape of the pressure that rose from the collapse. Because of the high degree of infinite pressure evolved and with breakage in the pressure built by the collapse and the pressure raised by it the strength of the expansion caused by the cosum collapse not only put everything in spin but also allowed only one way for the pressure to escape and that was to make concentrated. The young universe raised by density due to the raise in the pressure had no direction but to shrink backwards in on itself as it did when it collapsed. Nothing escaped the realm of the cosum bubble and its early mastoid.

Reviewing these two existing parts playing in the early universe's origin we for the second question have to answer the question about the speed of light. The speed of light seems to be an optical about the universe when it was first born to even today 13.8 billion years later in its history. The speed of light optical seems like it can't be reckoned with. Albert Einstein famous quote nothing can travel faster than light seems to useless to try and change its relativity. But the fact remains the speed of light at the earliest time of the universe's young origin during the expansion haven stretched and cause space to warp due to the flexibility maintained in relativity that classifies its character condensing of matter during this young time and birth of our universe space was not originally tough enough to withstand staying warped by the universe's creation. Space stretched and curved back on itself as relativity predicted in 1915 in general relativity theory says it did. To this effect space warped and shrunk back on itself allowing the physical nature of its flexibility as we know in today's modern physics. And by developing this way the speed of light haven condensed matter into the smallest of its extreme molecular structure formed. Everything we know as matter cannot travel faster than light. It is only that the warp age that space underwent doing creation that expanded and shrinkage that concentrated matter that space co-moving was warp enough and beyond its original fabrication allows for an open source speed limit.

What physicists do know about the wit of the universe and its creation that warp age of the vacuum fabric allows for a faster speed limit than that of co-moving planets, stars and even galaxies undergo even when they are set in motion by the wheel of inertia. From a subtle point and view space is motionless co-moving as a whole piece of the pie and everything in it is in motion due to an axis measurable by a rounded universal edge an edge whereas in a flat universe type doesn't exist. Event horizons as we observe them on the different planets like are planet earth are seen by the overview of the planets spin and is continuous. The universe shaped due to its existence inside a closed avenue of empty space was shaped with no means from the release of the built pressure created by its collapse nothing escapes nor was there any way for chemistry of its origin to escape out of the space barrier that surrounded it.

Everything in the origin of the universe was built by a greater force than what we perceive in the speed of light as a universal constant. The warp age that occurred during its creation is what allows the universe and all its matter to expand into as it is by the warp age that space is expanding from. Matter that reformed into a material mass that cannot escape the Einstein limit has nothing to do with the space fabric warped in the venture. Flexible to the degree a ship having the capacity to do so can travel faster than the celestial substance that doesn't attain a difference in its co-moving velocity.

The universe created from a single point as the standard theory explains it has been expanding and growing bigger since first sight of the anomaly. It is in this sense what allows that the galaxies are expanding away from each other and at a greater rate of speed than the velocity of light. As the universe expands the space expands along with it and according to the facts the space gravity is expanding faster in the forefront expanding edge we know as space.

It is space that even to this day and age of the universe that continues to fill in the cracks and curves it created in the beginning. By the weight the celestial planets, stars and galaxies tug boat through space by it is the space at the horizon end that is in expansion space that is moving faster than the matter that was formed in its way. From the single beginning origin in space the expanding universe widens the space between its galaxies that also widens the empty space between them. The further the object of the energy between two masses the least it is that any energy exist at all and is how travel at velocities faster than light can be weighed. (zg)

Hubble coordinates of the space foundation are the same that were created by Albert Einstein in the early twenty-first century. The factors for elevation in the modern physics science are led only by the acceptation in the rule. In a space field hence that is characterized as a repulsive carrier force there exist no acceptation that it is the force of space gravity that weakens the walls around the bubble that are universe is shaped by. It is the same force that created the cause and effect in the cosum' collapse in a closed universe where all the matter existing in it is too small of an amount for celestial gravitation causes to be formed for a universe it might expand forever.

According to Albert Einstein the speed of light was rectified when the universe was smaller than the point at the tip of your finger. As the universe grew it became more and wider between the galaxies. The speed of light changed also everything we thought we knew about the universe was figured wrong. The gravity mass as we knew it grew a part over time and the energy it possessed weakened. The speed of light was no longer measured in the manner it was long before centuries earlier because the matter the mass energy generated from only existed in the dead of space.

The Push of Space

In space, everything in open space is in momentum. Most physicists believe this universe's motion occurs because of planetary and galactic weight and is based on the facts that the universe itself doesn't rotate. But when we observe the idea and how everything is in a planetarially round shape spinning curvature is what makes the oval universe itself an exception? In outer space it is not all free fall coordinates that measures how fast a ship can travel at. It is divine intervention flight in zero-point gravity and how the vacuum is in nature. The deeper the area of the vacuum the faster a spacecraft might travel. There is most likely to exist deeper regions of space regions and planetary sections that free fall backed up by propulsion accelerations will push a ship to a most extreme velocity.

Over the decades the theory about gravity also researched by Johannes Kepler born in December 27, 1571 – November 15, 1630 who was a German mathematician, astronomer and astrologer who created the first model of the solar system which he was held in treason for by his ideas. When we review those facts about gravitation at the celestial scale amongst the planets and stars we have to look at the whole human picture. The universe spins slow dragging dark space with it as it rotates. As a whole it is the weight of the galactic matter that space forms into a rotating disk plate shapes and curves with. It creates a virtual fabric that allows the heavenly bodies to become impressed into in deep impressions in the fabric. It's like the universes spin forms a non-penetrable surface relativity calls the fabric of space when in reality it is the universes spin that forms this surface made grid. If the universe did not conform to this analogy dominion stars and suns would dominate space pulling in nearby spheres into its realm. But as it is the spin by the universe causes matter to push towards its outer rim or its edge. We look at the universe in allusive orbitration space pulling on the galaxies where in total motion it is the weight of the galactic matter that is slow dragged rippling a flat rigid space disk plate. Inside the galaxy's solar systems of stars and planets hardware retains an orbit due to the swift in the impression the suns retain on nearby planets but are contradicted by the universes spin. The density of a planet's atmosphere stretches into deeper realms in deeper regions inside the denser galaxies that exist throughout space. But the universes spin is what causes the alignment of the planet structure that lay inside revolving galaxies, nebulas and clusters of stars.

As the universe expands it gets bigger and getter creating a constant push on the galaxies shadowed around its core. This background radiated cosmic image is the solidity of a mass or inflating balloon as Erwin Hubble describes it hovers over the galaxies from inside its three-dimensional core. Its actual substance is unknown as this time but from reading this book I may have given you some ideas. To any degree it expands and has been expanding since the big bang or beginning of time as we know it.

As it gets bigger the galaxies behave smaller pods of interest but the galaxies are where all the stars and planets reside in and are constant taking in tension from the expansion. The universe is actually that blue round image I show on the cover of this book and inside chapter. The only difference is this sphere has all the galaxies embedded around it. When we look at the expansion we see the galaxies spreading apart from each other. Inside we observe the nothingness of empty space on one end and enlightened radiation one the other end. We are so far away in its distance from us besides being inside the Milky Way that we don't even realize its there. Its only because in the uniformity of the expanse space shows it to be in uniformity that we know it's there.

Einstein's view of our universe is just empty space. His theories embark on the idea that there does not exist a cause for distortion, space curvature and gravity waves and that they are all just an effect of the dark vacuum matter or space fabric but now you can see his observations are wrong. His idea that the distance it takes light to get to earth in an Alpha Centauri observation of eight minutes compared to local time here on earth that is so slow in a ship means time travel is possible is a unique one. Something outside or beyond reach only means it'll take longer.

As the core expands it gets bigger at a constant velocity that keeps its uniformity hold on planetary matter. That's good for us as living human beings. But it's that constant mobility that keeps things moving. The galaxies spinning, the planets and stars inside them and time to stay constant as we see it. If any of this failed and the core stopped for even just a second we would feel it. The tension the core holds on all planetary matter and the galaxies is what keeps everything working the way it does. No more no less. Without it we would be dead in the water as some say. The smaller anomalies like the tension of the earth's sun has on our planet would engage and we would seemingly fall towards that center singularity of our earth's sun. The fall would create chaos throughout the cosmos. It would be like the worst of earthquakes we have observed here on earth. It would be the end of life as we know it and or possibly if the universe's core were to slow down or speed up consequences will occur and dramatic ones.

These anomalies are the newest ventures about the anomaly of our universe. Chase to shove these realities are real ones. An invisible universe with no expectations is denying the truth. The factual basis that the cosmic background image is a reality is a unique acceptance for advance physics. With it we do not deny ourselves the truth. Research and evidence is a constant. One unique observation opens the doorway to the next and so forth there is no end. It's a compilation of information that leads us to further advance science and technology in the sciences, space and cosmic foundation.

If the universes planetary matter did not drag along its maneuvering the dark fabric the planets and stars making up the systems would fall downwards into the anchorage of great suns throughout the grid. It is the universes spin that the weight of matter traveling at terminal velocities some faster than others that create a balance in the universe all together.

On a subtle bases size mass of celestial spheres great and small heavy and gaseous are separated by 'similar' weight mass the spin of the universe creates, its energy mass only creates a shield for resistance amongst the bodies all matter is caught within the deep impressions of the dominate planetary realm we know as galactic celestials. Early physicists leaned down towards the idea that it was energy and attraction that the dominion of the gravity force manifested itself. But the truth of the matter is - like a moving car at is not friction that slows the carriage to stop but the lack of temporal force acting on it. Unless force acts as a continuum to moving objects like cars, boats etc.., they'll all slow from and cease to move.

Free fall in space and abroad earth surfaces venture gains by the lack of thrust. Terminal velocity acts as a balance between an objects 'weight' and its 'velocity'. As one free falls from the sky it is interrupted by the earth's rotation and tension of its orbit. Each second on the clock the earth moves 18.5 miles per second intervals per second a free falling object is pulled by the pressure elements (air, heat, wind rain) bestowed in the atmosphere becoming heavier and heavier as it reaches the surface. As it falls it's caught within the planets rotational track that layers the atmospheres invisible element forces twisting by the planets velocity. Bound by the planets elliptical path none of this changes to any much of a degree over time, the second units of a time clock interval seem to be a constant amongst everything that exists throughout space even the pulsar energy star impulse waves.

Space seems to be divided by elliptical plates that each section of space is divided by. The solar system divides its space area from a proceeding existing star system or system of rotation formed by some nearby system of planets outside our system dominion of planets. Elliptical highways are formed between these plate activities throughout the grid. There exist many systems in a galaxy each rotating plate traveling faster or slower than the other. The elliptical path between them act as highway passages for spacecraft flight this discovery of elliptical plates was finally discovered in 2013 by a satellite passing outside our solar system.

There exist two layers of space that exist in empty space field. The zero-point disk alignment the planets, stars and galaxies lay and impress themselves in bending the fabric's upper disk plate or elliptical disk plate as I explained about in the earlier paragraphs. Then there's the empty space field grade lying above the grid. It is the planets and stars that layer the space grid with impressed wells that lay deep within the top of the disk plate. Solar winds form at the dividing caps of curved space and the planet disk plate each celestial implants with its positional force of its weight and rotation. Above the impressed planetary disk plates, like the one our sun has impressed on the fabric galaxies have impressed the greater impressed grid which 'open space' lays above them if the fabric of space acts with the impression of galactic matter than it is not empty because the planets and stars act in its midst. But above this darken element fabric lays an open terrain of empty space that acts as free space and is not occupied by celestial matters.

Free fall velocities and terminal speeds measure at all points and areas of un-occupied space. The free fall of universe galactic planetary matter as it orbits allowing momentum between the grid and empty zero-point gravity (grid gravity) from which our solar systems rotate and orbit in and are formed caught inside deeper dense perimeters like in our Milky Way Galaxy.

History about the gravity force researched over the centuries have led scientist like Isaac Newton and Albert Einstein amongst only to mention a few to believe it is 'energy' or the 'attraction' of that energy inside matter that creates the gravity force. But is the impressed force of the universes action it takes amongst its celestial sphere and galaxies that its layered zero-point empty space element raises out from.

Neil Armstrong in 1963 headed an experiment on the moon by dropping a feather and a hammer at the same time as he sat sitting at his spacecraft's upper step form. He performed this experiment to challenge Einstein's theory about object mass and gravity. They both fell to the surface at the same speed. The hammer no faster than the feather but the fact remains that on the moon that retains as little as eleven per cent g-force that the lack of planet element existed on the moon's surface terrain. That eleven per cent of 1 g^2 comparison to earth's gravity equals about 1 per cent close to nothing. There existed no chemical elements as earth, air, wind or water on the moon. Its energy star content is near zero gravity as Armstrong could jump fifteen feet through the moons atmosphere only falling due to his measurable weight. What the experiment did prove was that the space that layered the moon's surface that retained no atmosphere at all – showed that space is a measurable zero-point gravity realm even if it surrounds a desolate satellite like the moon.

We look at warp drive as a means to make space travel faster than light based on the possibility that we can warp space to help make a ships maneuver through space to act in our favor. Kawecki Warp Drive looks upon space as a vacuum. A vacuum which the force of gravity like it reveals itself on earth doesn't. In space there exist two enterprises. The first is the fabric of space developed by Albert Einstein is the space matter that asserts itself on planetary matter and makes matter move in a particular way whereas in the presence of matter space bends.

Known as the fifth element the fabric of space is what mobilizes the planetary spheres and matter to shift into a constant momentum. Above from which the planetary stars and galaxies are mobilized is the surrounding vacuum. Gravity according to relativity is the activity of the material fabric and its activity consistency on matter. Around this activity is the vacuum which acts as the mobility and atmosphere that acts to form the third dimension the area of space that allows mobility between the planets and galaxies abroad. As a whole we see the universe in all three dimensions one as a straight line across two space allowing the mobility of up and down and third dimension as a global scale for the first, second and third dimensions to exist as one. What we know as the fourth dimension time is the character of Albert Einstein to enable the arithmetic to act above measuring velocity and distance. His insight made it assume the idea that time which acts as a coordinates to measuring distance and velocity he believed empowered a whole dimension by itself. He believed time was a function of the three dimensional universe and not the function of a measurement as it is proposed as being.

In physics there exist the explanation of hypothesis based on the quantity of being either positive or negative and that these two aspects can be placed in any position to characterize a foundation for the hypothesis. It can be positive with specific results negative with specific results both positive and negative and reversed to research the validity of the hypothesis or question. It is this type of terminology that the outcomes of the specific foundations of a theory is formulated and are taken into consideration with. These same equations are what are used in equating the arithmetic for positive and negative energy which formulate the functions and activity of the material universe and its gateways. The Alcubierre drive shifts space around an object so that the object would arrive at its destination faster than light would in normal space. We observe the equations in relativity and find that traveling up to a velocity close to light velocity is also possible under normal conditions. But what we want to look at is the ability to use relativity's equations of energy along with Alcubierre warp drive manipulations of space and advance our space journey even faster in normal space. We look at Einstein's equations and discover the fastest a mass can travel is 186,000 miles a second or the speed of light. We advance this speed limit with Alcubierre warp drive using Einstein's equations of light speed with Alcubierre warp drive and manipulate the fabric of space into allowing our ship to travel into the superluminal speed range. But even with Alcubierre warp drive theory it is shrunk away due to the lack of energy to surpass the universe's speed limit. We look at Kawecki Equilibrium warp drive and try and utilize the two methods of Einstein's equations and Alcubierre warp drive trying to discover a greater means to travel even faster than light measurable using $E=mc^2 (v^2)$ and the warping of the space fabric within our means. In Kawecki Warp drive or Equilibrium warp drive we re-invent the equations in relativity and recalculate the indifference in the Einstein equations based not with gravity and mass barriers where relativity places his barriers that creates a speed limit without reinventing the equation to meet with the new equations of the vacuum as well as equating numbers of smaller units within the seconds in his clocks. We review Einstein's equations that explain time dilation that acts accord with an action due to velocity but more so is also based on a clock in a gravity field. We review his facts on the constant speed limit based on infinite mass energy as a barrier but again his infinite mass barrier is weighed against an external force or inertia activity where in space such activity ceases to exist in the extremes as

G1 - but measure zero point in a vacuum. We also review Einstein's idea that time travel interferes when trying to travel faster than light radiation particles and put no efficient on the fact that time travel has not been proven to even exist at all and could well be a fable. A side from a few other ideas in relativity that I must say are viewed on particles that we as physicists can observe not taking into account those particles in this era that we cannot observe yet under a microscope where theories today are based on possible predictions are used as evidence. The fact that modern physics is also based on forward reverse ethical observations that theory the cosmos and theoretical fact with the idea that an antiparticle observation makes theoretical observation and theory a reverse prediction published a evidence maintained for the laws of the nature of the cosmos. In other words, we look at radiation or light in both a massive object and a mass less observation of something's we can't observe as being actually mass less at all even though light is rejected to enter through the fabric project ling effect of dark space. In Einstein's theory of relativity both special and general relativity he asserts a speed limit on the source of energy saying that the energy cannot become greater than that that can maneuver an object close to light speed. $E=mc^2$ calculates that mass and energy are stricken from exceeding a balance that he claims balances as an equivalence and are both limited to the velocity light travels in a second. Measurable of a mass less particle like the quantum particle having no mass Einstein admits a massive object in space weighs nothing in the comparison based on his speed of light constant. He also asserts a constant acceleration is necessary to even travel close to light speed yet we know that in space due to the vacuum this is not the case. That a ship might travel in impulse acceleration variations and remain at a constant speed without constant acceleration as relativity predicts. We also know that during the intervals of this type of constant acceleration that the ships engines can be recalibrated towards a greater acceleration speed that when engaged can increase the idle impulse velocity to advance into a greater faster coordinate's velocity. This of course could be made to advance the ship into unknown advance velocities over the course of a coordinated journey. We review Alcubierre warp drive he illustrates in forming the space fabric into a wave that pushes at the rear of the ship and can be used to accelerate his ship with no real bearing to the amount of energy needed to be engaged to travel faster than the speed of light. We again look at equilibrium warp drive by Rodney Kawecki and refer to the foundation of Alcubierre warp drive with the fact about our universe. That when observing our sun in the cosmos realm and space that if the sun were to implode, explode or suddenly diminish into nothing - the wave it would create would be enough to change the worlds around it in its space and cause possible death to the people on earth and possibly beyond the solar system. With this in mind we know for a fact that a cosmic equilibrium waves so great as for the sun create it would well be more than enough to push anything less great than itself towards a velocity that would exceed the speed of light - especially a starship measuring a piece of dust attached to a grain of sand. Equilibrium waves made on the fabric tension of a moving vessel can solely be based on the vessels velocity. Its source of energy creating the mobile force is what lays in question. We look at the vacuum zero point fields where due to an object mobility causes it tension where finally the tension changes into some sort of cosmic space wave. We know by comparison to predictable sun activity that the tension of such cosmic filler can be solely based on a ship velocity. That our sun's vanishing into nothing can be deferred into an amount of velocity. That a ships velocity will account for how much tense maneuverability will become. Since the ships mobility is measured by the amount of energy force its propulsion is acquired we need to account for our ships source for energy. We again review back with relativity and Einstein's great equation $E=mc^2$. The same way Einstein reviewed the facts for the barriers he created in his equation we as physicists' should look for ways these barriers can be unfolded. First we look at the space the vacuum a vacuum empty of gravity and measures zero to gravity. We can also look at gravity in space as being repulsive because gravity in space is pushing things outwards and away from the edge of space and from itself noting the galaxies abroad and Hubble Theory. We appeal the idea that gravity on earth is constant and that space acts as the dominate equilibrium of space where the planets and stars in galaxies are distortions in the whole of space as a universe by which all space acts first repulsive changing differently at the celestial scale. If we look at space in the same way, we review are earth - we can make assumptions that can unfold the barriers in Einstein's equations on space. Earth of course retains a gravity field measuring 9.8 m/s at free fall. In space gravity measures zero as a universe realm there exit little to no momentum in space. Mass in Einstein's equation measures also zero because there exists no energy in the space realm as it does in earth's realm atmosphere. (m=0) Where there exists no mass there can be no kinetic energy neither this is a fact of modern physics today. Since are ship measures mass less in space there exist no kinetic value for additional energy are ship seems free to maneuver through space at any speed its energy enables it too. Clocks speed up in space because of the lack of energy in the field where the same clock on earth ticks slower due to the distortion. Distortion causes everything in the distortion field to act differently than if a field lacking any energy exists. Einstein based time travel theory in relativity on the basis of what would happen if you could travel faster than light. After what you may have read in this book the possibility of time travel may not even exist at all. Where an Equilibrium Fabric warp accelerates a spacecraft pushing the vessel from behind having the time to build extreme warp drive energy or power in the ships energy is available or any distortion energy in the way that might slow the ships production as it travels through space a wave able to be formed with no real barrier shown to raise any type of limit the lack of distortion while are ship travels midfield deep space what turns out to be 186,000 miles a second due to the impressions explained in this new theory the positive energy from the negative distortion in the field are ship travels 372,000 miles a second with no account to ($E=vt^2$) whereas the faster the velocity the shorter time it takes to get there or in no time at all. The Universe cannot be solely based on a single second of time. No known physical particle travels faster than the speed of light but yet only a quantity of a force. `Instantaneous Space Travel' is the fastest anything can travel...particles are pushed in accelerators they don't travel at velocities close to light speed. I've written four books on faster than light and the origin of our universe and have discovered that even though particle matter doesn't seem to break the speed limit set by relativity really doesn't mean if it did time travel would be possible. The big bang was faster than light but didn't create any faster than light particles it most likely what they submerged mixture of particle matter. Why this is I have no clue. There seems to be a wall a head of any force that tries to assert itself against dark matter space mass element be it a mass of gravitons that form into a tension fabric or space projectile mass referring to the vacuum with a defense mechanism that changes form under quantities of pressure or force. This barrier of space tension in the presence of matter causes space to curl, bend and stretch over and around it - yet there exists nothing in literature about this cause and effect or reason for it. We look at the universe being able to bend, stretch and curve in the presents of matter than believe time travel is possible because of its potential. Wouldn't being able to travel faster than light mean experiencing these affects? If the universe was pushed into motion by an effect of transgressing how can one say that this transgression can be reversed?

Time travel is a myth spooky at a distance while matter sits on top the Hubble balloon which pressure seems to continue to expand from some place other than in the bubble itself. This pressure balloon is invisible in nature three dimensional in mass as galactic energies stretch through it across its three dimensional radius from galaxy across to galaxy energy forming a three dimensional gravity spin where at the center of the bubble mass the energy massively twists creating a twisting effect on the galactic bodies frantically shrunken into and impressed on the fabric on the top of the balloon surface fabric. Bound by the galactic density of the field as a whole a ship would travel above the fabric tension the galaxies form in a zero-point levitation gravities distance vacuum mass. Any mobility would cause an automatic free fall response into an endless zero-point vacuum directive by which nothing exists except empty space and which under these circumstances - time travel wouldn't exist at all. To be able for a mass to travel faster than light the mass must act as a conductor from which its frame or body is repulsive. In vacuum gravitons act on moving bodies no matter what their velocity or size. Graviton's push a positive mass away as a natural resistance in space the same way it makes planets rotate relative with their mass. The larger than mass body the slower the mass rotates over the fabric of space. This is because a mass after a specific point its weight compensates and acts against the graviton groupies that can no longer meet the requirements to push on the mass harder. The idea here is to be able to make graviton particle mass groupies to massively bombard a ship mass and to increase its repulsive push by attracting more graviton particle membrane towards our ship's body mass in the same way Alcubierre warp drive forms a wave with the fabric but in this case a graviton particle processor that attracts a multitude of graviton repulsive membrane to the rear of the ship mass. You can think of this as the speed that space-time reacts to small changes in the energy momentum content. Gravitons are the quanta of these ripples. You can argue that gravitons are mass less spin-2 fields. The mass less part means that they travel at the speed of light. If gravity is indeed the bending of space time are gravitons even needed? Consider the Earth orbits the Sun by constantly being attracted to it, yet it takes light from the Sun 8 minutes to get to Earth. Consider also that if a galaxy is 100,000 light years across it would take 100,000 years for light to get across it. Now consider that gravity is holding the galaxy together therefore making me conclude that gravity is instantaneously holding the galaxy together across the entire 100,000 light years of distance. So is gravity faster than the speed of light? We look at gravity as the fabric that holds matter in a specific place with no spoiler. It is the tension of the fabric that bends in the presents of the matter - graviton affects matter lifting the burden off the fabric element thus pushing it away gravitons are reinforced energy particles stimulating from a deeper origin inside the fabric balloon. Since the universe is made up mostly of larger planetary bodies called galaxies the interior between the galaxies through a cross 3D sectioning of the bubble is what twist the galactic masses in the same manner as a north pole pushed against another positive north pole will push both entities to curve.

The Universe Age and Circumference 3D Model

How big? They say nobody can actually determine the true age of the universe but Kawecki holds fast to the idea that time (t=0) is a good starting point.

Beginning as a small initial "seed"

T=0 + 42 miles a second Hubble Receding equation.

450,000 miles per second is the strength or density of acceleration the universe seems to be expanding at. This observation is created by calculating the diameter of the universe as made by relativity at 93 billion light years across.

450,000 m/s @ 42 m/s *hubble*

T (0) 42 m/s multiplied by 2 equals the sum between (T=0 and Hubble) universe expansion.

13.7 billion years old is the observed age of our universe according to specifics.

From T=0 to today 2020 yr.

Zero m/s to 42 m/s decrease of density from zero m/s to 42 m/s. over 13.7 billion years.

Earth mass

93,000 m/s speed of light E=m½c

93,000 m/s speed of light mass E=m½c all includes g

186,000 m/s speed of light E=m⁰c

186,000 m/s speed of light E=m⁰c

Total

Total: 372,000 m/s + 88,000 m/s = (450,000 m/s universes expansion rate of velocity)

Total: 450,000 m/s.

42 m/s Hubble expansion

+ 42 m/s Hubble expansion rate @ 13.7 billion years.

= 84 % of 100 (88 m/s)

Of a hundred percent 450,000 m/s is equal to 88% of 100 in relationship to the age of the universe.

100% minus 88% = 12% of 100

450,000 m/s X 12% = 54.0

450,000 m/s + 42 m/s = 492,000 m/s (@ 500% or ½)

42 m/s

500,000 m/s minus 450,000 m/s = +50,000 m/s

13.7 billion years @ Flat (2D) universe (m) mass = 33.3 % TODAY

26 billion years @ round (3D) (13 billion years + 13 billion years = 26 billion years). @ pi)

(m) mass @ 33.3 % = 66.6 %

13 billion years + 13 billion years + 13 billion years addressing pi (mass)(33.3)

= 40 billion years @ 99.999 percent.

Results: using approximate numerology

Flat Universe = Age is 13.7 billion years old

Round or Oval Universe = Age is 40 billion years.

Percentages

450,000m/s + 44 m/s = 494,000 m/s (@ 500% or ½)

42 m/s(@500% or ½) space field density

or ½) space element density. TODAY

Plus, ½) space field density future

= universe expansion @ approx. 1000,000 m/s. (=100% percent)² halfway mark

Based on these equations the galaxies will become too far apart for journey in about 10 billion years in the future.

The Universe according to Planck

0.00000000000000000000000000000000000000000054

5.4×10^{-44}

Planck Time (in seconds), the shortest meaningful interval of time, and the earliest time the known universe can be measured from.

0.000000000000000000000000000000000001616

1.616×10^{-35}

Planck Length (in meters), the size of a hypothetical string. Lengths smaller than this are considered not make any physical sense in our current understanding of physics

0.000000000000000000000000000000911

9.11×10^{-31}

Approximate mass (in kilograms) of a stationary electron

0.000000000000000000000000001

1×10^{-27}

Approximate density (in kg/meter3) of the universe as a whole

0.000000000000000000000000001673

1.673×10^{-27}

Approximate mass (in kilograms) of a proton

0.000000000000000000000000001675

1.675×10^{-27}

Approximate mass (in kilograms) of a neutron

0.000000000000000000000000005

5×10^{-27}

Estimated critical density (in kg/meter3) of the universe, to allow a steady state between expansion and contraction (about 5×10^{-30} g/cm³)

The Cosmic Back-Ground Universe

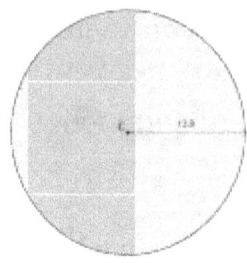

3

Circumference model = $2\pi * (-1/2 \text{ m})$ = 42 billion years

Defining Free-fall

A freely falling object has weight W=mg, where W-weight, m-mass of the object and g-acceleration produced due to the earth's gravity. An object kept in a lift which falls freely, weighs zero on the weighing machine, but its actual weight is still mg.

If accelerated by an external propulsion rocket or device, acceleration takes on the magnitude of the objects free fall status. The faster the object is falling the greater acceleration is necessary. Accelerating out of its normal gravity free falling, and taking on acceleration does not mean it's taking on energy-mass. The object energy-mass does not rise and fall as some scientist thought earlier in the century. It is the curvature or change of acceleration that causes a higher degree of acceleration and propulsion to have to be applied. Flight is not free. Nor does it change the weight (mass) of the object.

Does weight affect gravity or how fast an object falls?

Both objects fall at the same speed. Mass does not affect the speed of falling objects, assuming there is only gravity acting on it and that gravity is the same.

Why is an object in free fall weightless?

They are weightless because there is no external contact force pushing or pulling upon their body. In each case, gravity is the only force acting upon their body. The force of gravity is the only force acting upon their body. The astronauts are in free-fall.

Do heavier objects fall at the same rate as lighter objects?

In real life, heavier objects sometimes fall faster than light objects, but not because of gravity. Gravity makes all objects increase their speed at the same rate, regardless of how big they are. If gravity didn't exist they would fall one faster than the other.

Does zero gravity exist?

The sensation of weightlessness, or zero gravity, happens when the effects of gravity are not felt. Technically speaking, gravity does exist everywhere in the universe because it is defined as the force that attracts two bodies to each other. But astronauts in space usually do not feel its effects.

At what height do you become weightless?

Any object that is falling freely is weightless, no matter where it happens to be. This can be the International Space Station at a height of 200 miles.

Is gravity repulsive or attractive?

Both in the Newton theory of gravitation and in the General Theory of Relativity the gravitational force is exclusively attractive one. However, the quantization of gravity shows that the gravitational forces can also be repulsive.

This idea has been argued because to be an attraction force, one object has to be positive and the other negative but science shows us that the earth which is positive matter and a falling or accelerated object is also positive matter therefore 'repel'.

Is gravity a law or a theory?

This is a law because it describes the force but makes no attempt to explain how the force works. A theory is an explanation of a natural phenomenon. Einstein's General Theory of Relativity explains how gravity works by describing gravity as the effect of curvature of four dimensional spacetime.

Can we explain gravity?

The answer is gravity: an invisible force that pulls objects toward each other. Earth's gravity is what keeps you on the ground and what make things fall.

Is gravity a real force?

In general relativity, gravity is a fictitious force. In classical mechanics, fictitious forces are not considered "real" forces. If you believe that inertial forces are forces, then gravity is a force. If you believe that inertial forces are not forces, then gravity is not a force.

Gravity is the force that governs a closed system like planet earth. The earth acts on its own potential inside its own atmosphere. While a planet is inside the universal realm and its space its celestial activity is measured in contrary.

Earth g

Freefall….is a free action that relies on two forces. m (m1 and m2) forming a single repelling force. Free fall occurs when object A earth, and object B the mobile object is equal in mass allows the smaller surface mass ton fall freely. {red 'm' is for matter by itself without mass (energy) as relativity implies, when in space mass is zero implies}. Energy pertains to each mass.

A force of 'g' that is at rest acts as 9.8 m/s when it's at rest, with the primary a mass earth.

Acceleration occurs the rest force is asserted to the acceleration force on mass B. the object that was at rest.

Interaction with mass A. earth is null when acceleration occurs. The force of acceleration only applies with the moving force object B, g is still inversely implied on mass B (as its weight. And direction and distance is applied to object B. and its momentum.

There exists no real change in the mass B, weight, (9.8 m/s) which equals 1/10th of one hundred percent meters a second.

Mass is the interaction between moments of acceleration that occur in the moment and is a deterrent to free fall. it is what keeps things together as a single unit.

Net force towards destination is applied until a totality of acceleration is reached.

It is not possible to feel speed while in a spacecraft. Astronauts in orbit travel at 28000 km/h but feel absolutely nothing, even if they're outside. You feel the speed only if traveling through the air, where you feel the air dragging at you. Speed does not cause any harm at all, as you never feel any of its effects.

It takes place in the not-so-distant future all right here in our own solar system. There are no pew-pew lasers or faster-than-light space travel. When humans are on a spacecraft, they either "float" around or use magnetic boots (except when the spacecraft is accelerating). There are no "inertial dampeners" in the Expanse. Not only that, but it has interesting characters and a compelling plot. I like it.

The "Secret" of rocket propulsion for space travel.

If A exerts a force on B, then B exerts an equal and opposite force on A. Or, in the case of space travel, if a mass (m) of fuel is pushed out the exhaust of a rocket, then the rocket will accelerate in the opposite direction the direction the exhaust fuel went.

Kawecki laterally opens the door to define the gravity force on the foundation of nature not energy and deciphers the equations and measurements that precede his theory. Why does falling objects no matter its initial weight or size all fall at a specific speed towards the earth's planet surface? What causes the action that precedes free fall and by what means does it happen? And finally – if the gravity force celestial or inner planetary atmosphere is not measured by the cause and effect of energy or energy mass that repels than what is it? In this book he will answer all these unknown questions about "The Path of Gravity".

When we talk about gravity we are not talking about the

fundamentals of energy and mass but more so is the activity of zero-point energy presence and surface weight. Space is different. We are talking about how similar or like material entities called celestials such as planets and stars repel forming the distance between them in space present on the fabric of space that impresses them into deep holes. In space yes it's the presence of planetary energy in the elements that commands the divide. But in the interior elements of a planet's atmosphere there exists the 'ladder of resistance' or an entanglement of free fall of an objects weight and acceleration as it falls. In the thinner upper atmosphere an object in free fall gains acceleration in layers through the atmosphere as the planet's atmosphere pressure increases making the object faster at variable speeds as it follows a curving free fall pattern towards the planet ground surface. Nothing travels straight down to the earth's surface.

The problem I have with ""G"' is in space similar matter (planets) according to physics repel and do not attract as today's modern physicists have been lead to believe over the years. Like a magnetic north pole and another north magnetic pole...they repel. According to the big bang theory if all was in advance a singularity bang universal, galactic or otherwise then these similar worlds of common ancestry matter ...*repel*... and don't attract wouldn't you agree?

If this is the case than - a planetary gravity field "TORUS" otherwise...pushes away objects as in universe expansion theory explaining the pushing away expansion. Gravity thus on earth in this sense free falls at 9.8 meters per second compared with 18.5 miles per second it's rotational orbit...this is a free fall elemental falling object wouldn't you agree?

Do wormholes exist or can they be made using the energy capacity of a star? Even a star cannot penetrate the fabric planets and galaxy's themselves shelve themselves on in space. To think that breaking through the fabric of space would be possible lays the end of the wormhole travel short-cut theory in special relativity.

It makes one believe whether or not Albert Einstein really believed that such things were special at all?

Rodney Kawecki tries to answer all these questions about gravity and more. This book will enlighten and supersede the most interested reader of advance cosmogony. Gravitation has at reach or range to infinity. However, it is the weakest of the fundamental forces. The gravitational strength is only 6*10-39 of the strength of the strongest nuclear force. In comparison we observe space on the scale of 5 % percent the volume and space the other 95 % percent so I ask you – where does the advantage lie?

Note: 10-39 equals 1/1039, where 1039 is 1 followed by 39 zeros. That is a very small number.

The strength of the gravitational force decreases as the square of the separation between two objects, as does the electromagnetic force. Although the gravitational force is much smaller than the other fundamental forces, it's impact concerns objects of large mass, such as planets and stars. Gravitation is what keeps the Earth and other planets in orbit around the Sun, as well as the Universe.

$$(\text{Lift-Vector (repel)} + C^2 = G^3) 3D.$$

The universe is expanding and we as explorers can travel just as fast as it expands

The universe has an expanding runaway gravity force with a velocity equal to 372,000 m/s plus 78,000 m/s or to be more exact in numbers: 450,000 miles per second is the accelerating speed of gravity as a carrier force that continues to grow greater and greater over time. All this in accord with the common standard theory about the universe as we know it there are two different origins about the position of the universe. The first is its matter oscillates inside an expanding bubble formation. The second is that matter oscillates outside the inflation of growing pressure. We will be deciphering the universe oscillating outside the bubble whereas the standard theories about the universe all maintain their sufficient is operating inside the bubble formation. This indifference between the two universe theories of sufficiency and universe structures science have only researched using the inside bubble theory over the decades.

Everything in the origin of the universe was built by a greater force than what we perceive in the speed of light as a universal constant. The warp age that occurred during its creation is what allows the universe and all its matter to expand into as it is by the warp age that space is expanding from. Matter that reformed into a material mass that cannot escape the Einstein limit has nothing to do with the space fabric warped in the venture. Flexible to the degree a ship having the capacity to do so can travel faster than the celestial substance that doesn't attain a difference in its co-moving velocity.

The universe created from a single point as the standard theory explains it has been expanding and growing bigger since first sight of the anomaly. It is in this sense what allows that the galaxies are expanding away from each other and at a greater rate of speed than the velocity of light. As the universe expands the space expands along with it and according to the facts the space gravity is expanding faster in the forefront expanding edge we know as space.

Reverse vacuum gravity is a review about space that allows a concentrated mass, like the balloon hypothesis about an expanding universe to supersede itself. Unlike a false vacuum the fate of the universe is not observed in the final hours. It slowly redeems itself backwards until the universe falls again. A time when all the forces we believe that hold the universe together diminish and are a collection at the bottom of the most gigantic gravity well emerges.

The universe is expanding and we as explorers can travel just as fast as it expands if not faster. In an expanding universe a ship has the ability to catch up with the universes expansion time variations by traveling at speeds solely relative to the universes space velocity. What this means is that a ship traveling at close to the gravity length of space expansion can travel to far away stars or galaxies that are expanding out of reach with time. We look at open space with the idea that the universe and all matter in it is expanding at a specific velocity. When observing tis time coordinates we discover that if we were to travel in the direction of the expansion we will be traveling towards the ends of time. But if we were to travel towards the sooner lengths of the expansion we would be traveling towards the early born time of the expansion thus towards a place in time when the expansion is hardly getting started.

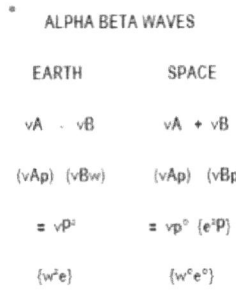

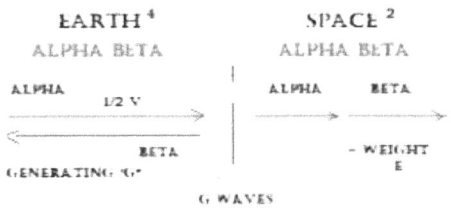

Making Waves of Light

The Push to the Chase

A ship in space that travels at light speed generates more propulsion energy than on earth that would otherwise be used for speed but rather it slows the object down. In reality speed is decreased by mass that would otherwise be used to velocity.

$$\{p = m°p^2/v\} + \tfrac{1}{2}F\}$$

The speed of light is a comparative with the understanding that we are talking about 'light' and not energy. That the content of how light has a velocity can be used to generate propulsion and velocity for space travel. This idea by no means indicates that light is energy or and can be used for anything but what it is used for. But where a comparative expresses regularity between the substances they can be weighed as comparable and therefore value. If you were to say a frog leap's a comparative can also be said that the frog jumps. There is substance in the words used to mandate the affect.

In the same sense that you need two people to create reason is the same idea that Newton shows that it takes two planets to resolve their distances from each other. The idea that if it was just oneself to decide something and nobody else was present your knowledge would be single minded.

In The Account To Re-write History in Physics

This audio is for the scientist to whom makes wishes for the progress of traveling to the stars. For people like himself this analog recording is just that a tribute to the landlords of universal survival and human networking.

The First Topal,

It has been in the greater good that new history applications and theories have rarely been implied in our history books. These world historians especially in advance physics have been accredited for their work so dearly because their work in these categories in science has been for the greater good of mankind.

These historians by name like Aristotle, Galileo, Isaac Newton, Albert Einstein and Erwin Hubble have all accredited scientific fore-knowledge of the greater good. Independently, these historians have made contributions to science that has led the world good on a path of accomplishments that have placed in lead to the world emperors throughout our planet.

All makes sense when it is applied in natural formation to the degree that specific insight is a pledge to greater mankind and future civilization around the globe. Such leadership is in the literature of the sciences, geographic phenomena map's, aspiration in economics, mathematics, historical discovery just to name a few - has re-led our species to succession. It is and has our goal as human being of a worldwide planet to adventure the stars. Its beautiful ness has blemished in the fairy tales of all world historians worldwide.

Traveling to the stars is a frontier conquest of nations that seems to supersede ambition. It is the cradle of mankind's succession to achieve greatest amongst men. This ambition in its right has blemished in the world of mathematics, physics, and natural sciences and biologic - behavior over the decades that supersedes even itself. It holds evidence in the plans of the tomorrow. A place nobody has ventured too yet creates the gateway.

We talk about conquest as rightful owners of antiquity, that hold the keys to the tomorrow of which we gather around.

Science has been one of those journeys. The idea we can reach out to the stars and travel there tomorrow is the greatest thought of modern history. It entails - progress for the days ahead of us, as we can mark our maps that have al - ready been written. We speak about faster than light space travel as a thing of science fiction until someone puts the right design on our table of tomorrow. Our inner thoughts about this science fiction fairy tales begin to build into a monument of discussion and some one believes he or she can built a ship. They look out at the sky in the late evening of the day and paddle the stars. But such an advance - would take possibly decades or at the least years to promote.

Like everything - that which has been speculated by earlier suggestions is in reality the format for the new discoveries of tomorrow. When Albert Einstein published his work on relativity saying the universe was flat dimensionally, it didn't stop Erwin Hubble from looking through his telescope. Nor did Galileo stop sir Isaac Newton from re-inventing a new telescope more refined than Galileo's lenses. Albert Einstein didn't stop researching Isaac Newton's analogies about a better lens that he discovered flaws in not Isaac's flawed lensing but distortions in space amongst the celestials. Einstein's venture to refine antiquity from what he viewed in Newton's ancient-al records of accounts, well you know the story.

The theory viewed in the hypothesis of the publishing of this book relies with the idea that our universe is three dimensional. Addressed in Edwin Hubble's theories he viewed the universe in three dimensions. The first of course is flat as Albert Einstein believed the second is two dimensional meaning the sight of things ran one way then turns. The third dimension was that the universe was round. That the galaxies were running away from each other at a great velocity and hiding behind an event horizon was the presence of a round oval ball-like structure that the galaxies laid upon.

Of course, Albert Einstein's quote was: If a ship started out on a journey in a specific straight path it would wind up right back in the place it started from". Of course, this could never happen on a flat surface universe as he thought. But in a-round type universe it could.

In the Hubble Universe this is what Hubble suggested. In Quanta Physics Theory and in the writing of this book the same analogy is respected. The only difference is though the physical universe is separating or expanding it is the space that is expanding causing the effect of expansion. It is also theorized that at the same time of the big bang as it occurred middle space became concentrated in the same manner as matter assumed to shrink after expansion from the bang. Space as the place of this revival and creation of the universe space shrunk and became concentrated and is what is the cause and effect of the universes expansion.

We all look at the universe in three dimensions in its physical formation with invisible forces acting throughout it as subatomic particles of all degrees pass through, around and crash into matter inside it and outside. As a whole the universe is a structure of energy masses to its highest degree. We view the universe as all these energetic forces surround us throughout space. Most of which acting at subatomic scales we cannot see or hear their passing. Yet we view their presence and with finite eyes know they exist in a reality we only assume exist yet assumes in our intelligence as a structure of true reality throughout the universe.

> The universe is expanding and we as explorers can travel just as fast as it expands if not faster.

The second Topal,

It has an expanding runaway gravity force - with a velocity equal to 372,000 miles per second plus 78,000 miles per second or to be more exact in numbers: 450,000 miles per second is the accelerating speed of gravity as a carrier force that continues to grow greater and greater over time. An accelerating gravity means that what is known as 'universal gravity' or space of the cosmos is a force created by momentum of a physical deity in space that because of its actions creates a motion in the space fabric. Unlike planet gravity that is the density of the planet's atmosphere that governs moving objects or movable objects on its surface since space is not considered to have a dominion planet as such and where space gravity is defined by Albert Einstein as affects created by the space element and its connection with matter, the space vacuum with thanks to Erwin Hubble is now defined in the shadows of the cosmic background image that distinctly shows that an expanding sphere a lurks in the dark vacuum of space. It is this mysterious sphere that gravity in space is directed from and created by. Its expansion keeps a constant acceleration of 'growth' or acceleration of size mass that creates the anomaly for universal gravity in space. It also is characterized by the vacuum density that measures zero because it's an open source phenomena. We can also postulate that universe expanse theory can account for not only motion throughout the cosmos but also in human and animal life forms.

All this in accord with the common standard theory about the universe as we know it there are two different origins about the position of the universe. The first is its matter oscillates inside an expanding bubble formation. The second is that matter oscillates outside the inflation of growing pressure. We will be deciphering the universe oscillating outside the bubble whereas the standard theories about the universe all maintain their sufficient is operating inside the bubble formation. This indifference between the two universe theories of sufficiency and universe structures science have only researched using the inside bubble theory over the decades.

Everything in the origin of the universe was built by a greater force than what we perceive in the speed of light as a universal constant. The warp age that occurred during its creation is what allows the universe and all its matter to expand into as it is by the warp age that space is expanding from. Matter that reformed into a material mass that cannot escape the Einstein limit has nothing to do with the space fabric warped in the venture. Flexible to the degree a ship having the capacity to do so can travel faster than the celestial substance that doesn't attain a difference in its co-moving velocity.

The universe created from a single point as the standard theory explains it has been expanding and growing bigger since first sight of the anomaly. It is in this sense what allows that the galaxies are expanding away from each other and at a greater rate of speed than the velocity of light. As the universe expands the space expands along with it and according to the facts the space gravity is expanding faster in the forefront expanding edge we know as space.

In reality, the universe is structured by the cosmic background imagery. An image that in reality is so far in distance, a mass in the center of the universe that physically stabilizes everything and I mean every throughout the cosmos. It is the center mass of the universe that after the bang formed at the center of re-formation and slowly started creating the motion throughout space the galaxies anomaly the motion of the planets and stars and system of stars inside the universe.

So what I'm I talking about when I speak about the cosmic core. What is radiation? In physics it is the emission of energy as electromagnetic waves or as moving subatomic particles, especially high-energy particles which cause ionization.

In physics, radiation is the emission or transmission of energy in the form of waves or particles through space or through a material medium. This includes: electromagnetic radiation, such as radio waves, microwaves, infrared, visible light, ultraviolet, x-rays, and gamma radiation. Radiation damages the cells that make up the human body. Low levels of radiation are not dangerous, but medium levels can lead to sickness, headaches, vomiting and a fever. High levels can kill you by causing damage to your internal organs. So accordingly this sphere is hidden from us by what distance.

Reverse vacuum gravity is a review about space that allows a concentrated mass, like the balloon hypothesis about an expanding universe to supersede itself. Unlike a false vacuum the fate of the universe is not observed in the final hours. It slowly redeems itself backwards until the universe falls again. A time when all the forces we believe that hold the universe together diminish and are a collection at the bottom of the most gigantic gravity well emerges.

The universe is expanding and we as explorers can travel just as fast as it expands if not faster. In an expanding universe a ship has the ability to catch up with the universes expansion time variations by traveling at speeds solely relative to the universes space velocity. What this means is that a ship traveling at close to the gravity length of space expansion can travel to far away stars or galaxies that are expanding out of reach with time. We look at open space with the idea that the universe and all matter in it is expanding at a specific velocity. When observing is time coordinates we discover that if we were to travel in the direction of the expansion we will be traveling towards the ends of time. But if we were to travel towards the sooner lengths of the expansion we would be traveling towards the early born time of the expansion thus towards a place in time when the expansion is hardly getting started.

The strength of the gravitational force decreases as the square of the separation between two objects, as does the electromagnetic force. Although the gravitational force is much smaller than the other fundamental forces, it's impact concerns objects of large mass, such as planets and stars. Gravitation is what keeps the Earth and other planets in orbit around the Sun, as well as the Universe.

Rodney Kawecki tries to answer all these questions about all space and more.

This book will enlighten and supersede the most interested reader of advance cosmogony. Gravitation has at reach or range to infinity. However, it is the weakest of the fundamental forces. The gravitational strength is only $6*10-39$ of the strength of the strongest nuclear force. In comparison we observe space on the scale of 5 % percent the volume and space the other 95 % percent so I ask you – where does the advantage lie

The problem I have with ""G"' is in space similar matter (planets) according to physics repel and do not attract as today's modern physicists have been lead to believe over the years. Like a magnetic north pole and another north magnetic pole - they repel. According to the big bang theory if all was in advance a singularity bang universal, galactic or otherwise then these similar worlds of common ancestry matter …repel… and don't attract wouldn't you agree?

As the core expands it gets bigger at a constant velocity that keeps its uniformity hold on planetary matter. That's good for us as living human beings. But it's that constant mobility that keeps things moving. The galaxies spinning, the planets and stars inside them and time to stay constant as we see it. If any of this failed and the core stopped for even just a second we would feel it. The tension the core holds on all planetary matter and the galaxies is what keeps everything working the way it does. No more no less. Without it we would be dead in the water as some say. The smaller anomalies like the tension of the earth's sun has on our planet would engage and we would seemingly fall towards that center singularity of our earth's sun. The fall would create chaos throughout the cosmos. It would be like the worst of earthquakes we have observed here on earth. It would be the end of life as we know it and or possibly if the universe's core were to slow down or speed up consequences will occur and dramatic ones.

As the universe expands it gets bigger and getter creating a constant push on the galaxies shadowed around its core. This background radiated cosmic image is the solidity of a mass or inflating balloon as Erwin Hubble describes it hovers over the galaxies from inside its three-dimensional core.

Is time travel possible? Some think it is - but really is it? Fourth-dimension a 'demeanor' measures by the threads of time as a dimension of itself. But space-time is based on the reality of the universes motion which is real and sits independent of it being it is the universe itself. To say time is an actual dimension that interchanges within itself past, present and future actually means time is absolute and nothing in it can change. So time travel is impossible. (It either is (present) was (past) or never will be (future) because these aspects about reality and the universe are absolute. They are either present or they are not. Time travel would then have to come from the future earth and it doesn't exist at all except theoretically.

Space travel is how fast gravity will allow you to travel. On earth gravity is 9.8 meters a second. In space the speed of expansion is 450,000 miles per second that's quite a lot. Expansion defines the universe's gravity. We can travel just as fast as its gravity force does. But where we are limited to light speed, a car on earth only travels half its intended speed because of gravity. In space we can travel twice the velocity because gravity acts as a push force. It is the cosmic cores acceleration and push rate multiplied by the Hubble constant, divided by the universe's diameter. That velocity divided by light speed equals 2.7 c.

If light speed is the constant in relativity than in space we can travel lightspeed. We travel twice the velocity of light without gravity's pull.

Rodney Kawecki has written several books about space. He has developed his new theory for the sole purpose to advance modern physics re-adjusting any flaws that may exist and re-inventing the cosmos and light speed theory.

Furturistic Ideas

        Having the Ability to Travel Twice Light Speed

The Third Topal.

Sooooo, we'll call that negative

Mankind would like to believe they could zoom through space at any velocity they could but only time will tell us when that day will come. Meanwhile we have to look at space travel in the same manner we did transportation here on earth. We have to wait until this ambition becomes a reality. It took us over sixty five years to jump from wagon wheels to automobiles. And I think traveling space is a bigger jump than cars were in the early nineteen hundreds.

Some believe we have to discover a way to explore the cosmos for different reasons. Some think we are destroying the earth's atmosphere with are early luxury gadgets planes, trains and automobiles but that's here on our planet. Space is a lot different. For one thing venturing the cosmos takes a gigantic turn on sufficiency. One mistake in space and there's no turning back. You can't just hatch a ride from someone passing bye because you'll be the only out there. Fixing any problem isn't easy and you're too far away from civilization to call for a ride. Though those are the possible fixes from being stranded in space they aren't pretty ones.

No space is far out there and sufficiency is your best friend. For right now we don't have the means for such space journey's even though we push the advancements of sending gate life's out into space for planet communications, devices and cable television it's not the same when you're sending human beings out there to explore. Robots and road drivers are expensive enough.

Technology only takes us so far. Manufacturing product and good test flights and success rates are great substance for achievement. Taking to advocate faster than light speed space travel is a big jump from small earth vehicles like cars and motorcycles. Light speed technology specifically for space intervention has been going on for only a mirror hundred years. Besides science fiction real space travel has been on the shelf when it comes to achieving progresses. Trying to invent logical space flight vehicles or rocket ships is no longer science fiction to the earth civilization. The world is realizing progress to exploring the stars is an advantage to civilization and discovering new places for colonizing. The planet earth is getting to be a small place when you think about all the people living on it. After all it's not a luxury item.

No the planet is real and so are its people. Trying to find a second earth isn't easy. And the distance for such advancement is astronomical. Light speed space travel is its only recourse for mankind's future.

What am I talking about when I say in space we travel twice as fast as we do on earth?

This might sound strange to you if you haven't explored space physics if you have your way a head of the game. The idea about space and space travel is the anomaly that space is vacant matter. Matter is the accumulation of matter planets stars and galaxies that over the eons have accumulated into small singularities from the big bang event.

Because space is vacant or empty an vacuum it is the distant between these singularities that patch the way for space travel. Having acknowledged the deity about advance physics you understand me when I say light speed is the fastest a vessel can travel through the cosmos. Have you ever wondered why today's modern space craft's launched into space travel at 22,000 miles an hour? That's quite fast according to local speeds here on earth but they do. The answer lies on the fact that engine mechanics and engineering has been mapped in the same manner that allows us as a civilization to explore the stars.

Mechanic engineering designed propulsion to agree with the cosmos. It lies on the fact that it takes two pounds of energy to make a one pound mass or object move. But what does that mean with space? It means that engine wise to make a ship in space move faster than light has already been acknowledged in the blue-prints. A two-cycle motor will make a one pound mass move on a terrain the same Isaac Newton's ball rolled down a hill. Two pounds of energy will make a one pound move and at a constant speed if the two pounds of energy is applied constantly like Newton's ground run per second per second in the same way the universe's background sphere advances per second per second as it evolves and grow bigger and bigger. This makes the galaxies move a part more and more over time.

For us though we need a spacecraft to maneuver through space and that takes an engine. This engine has already been designed but its capacity is what stands in our way to progress. We have to be able to travel light speed to manifest a star ship. Stars and suns have a lot more energy than the speed of light needs to make its rounds to twist its alpha beta particles into waves. A maneuver science has yet to manifest in physics but to any degree I will talk about it here.

D.C. motors maneuver the same way as a gas motor pushing cylinders at about 900 revolutions. It pushes one cylinder up and one cylinder down simultaneously to create motion. The same logic is needed in space. But a more refined theory. Anyways this engineering logic tells us we need two pounds or tons of energy for propulsion to move our 68 ton ship in space.

On earth a car uses two times the pounds of energy to accelerate a car half the velocity. What this means is on earth because of gravity slowing are car down - to travel 50 miles an hour it would take one hundred pounds of energy or propulsion. That's how cars on earth operate. 50 miles per hour takes 100 pounds of energy - propulsion.

When in space, the zero point gravity realm. Positive advancements are realm by positive responses. Negative responses are governed by negative résistance and negative force playmatics.

Since positive responses are related to physical engagements, negative engagements at below zero point energy (G) in space, retain negative responses.

Negative energy playmatics is the activity of negative force pressures in space only able to interact with negative playmatic responses.

Negative energy engagements applied as a force can apply with other negative playmatic responses but never with positive responses.

Example: A positive energy blast directed to a negative response carrier force will absorb the blast not engage it.

Positive playmatics in zero vacuum space is directly related to positive interactions. No resolution will occur applying a negative playmatic response with a negative one. A carrier force occurs connecting the force plays when they connect one positive one negative ground port.

Like I said the specifications have already been in the blue - prints for space travel and assist. But in space we don't have gravity acting on our body. Space is vacant of attraction and interactions unless physically applied. in fact space is a repulse environment. The ship sits idle in space because a constant pressure of two point seven light velocity pounds of energy - pressure creates a sitter position for all matter in outer space and that includes earth.

So to the cut of the chase are ship has an engine that can travel light speed. Light speed is our target in space engineering because the distance between the stars is so great it would take that amount of energy to get anywhere in space. Short runs like the moon and Neptune are local project planets that we work with today but real space is the anomaly.

Are ship has the capacity to ignite into a warp drive calibrate at light speed. The engines have been post designed with the capacity to move are ship and hopefully to light speed.

As we look at the design we understand that it takes two pounds of energy to maneuver a one pound mass our ship is a mass but it's not coordinated to a second mass that would induce a gravitonic vicinity. A lone in space it's not energy that makes it sit idling it is pressure. To create any real balance within a system of stars it takes at least two planets. Our ship is a lone.

It has the capacity to maneuver into light velocity over time but its engineering design places the ship with twice the energy modul-alities that because no gravity is around in vacant space and it resides mid field space and not too close to any planet or star - when it ignites and goes into acceleration mode are twin cycle engine and because no gravity effects on its body will venture from a single light speed maneuver change into twice light speed.

THE COSMIC CORE

Too look at it more likely. Because propulsion values are designed in the manner they are today a car uses twice the energy needed for acceleration to travel at a specific speed. That slowdown is caused by earth's gravity force. In space at zero point gravity the friction on earth that acts on a moving body and slows its progress acceleration rate in space is added to travel velocity. It is in this way a ship in space will travel twice light speed velocity using two cycles of energy for each pound weight of a ship that in space becomes vacant of gravitation.

Earth: Fe - ½v = m.p.h.

Space: Fe = Va = miles per {second, minute, or hour}

Square Route to the Stars...

Start with what you know: $a^2 + b^2 = c^2$

Put in what we know: $52e + 122e = c^2$

Calculate squares: $25 + 144 = c^2$

$25e + 144e = 169e$: $169e = c^2$

Swap sides: $c^2 = 169e$

Square root of both sides: $c = \sqrt{169e}$

Calculate: $c = 13e$ {d} *path of* c^2

Path of 'c' {'c' "C"}

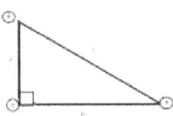

9.8 meters a second 'g'

1 mi/s = 0.0000000033356 c

{9.8 m^e/s}^{1} 10 mi/s = 0.000000033356 c

9.8 m^e/s = 1/10 m/s

speed of light = meters a second ÷ 299,792,458

1 c = 299,792,458 mi/s

Relativity makes mistakes

Assuming relativity is right and earth gravity or any gravity force, is an attraction force as it claims and this attraction is what causes a moving mass to be slowed down - than his theory projects the negative positive connection. But, if not - and the 'universal gravity force' is a repulse force there exist an indifference in celestial and space. Relativity claims this is not true as it creates an inter-relationship with planet gravity and space gravitation. ('G' 'g')

We should assume all gravity force interactions are positive as science intended it. Relativity has violated this compromise. We must also assume that any and all gravity forces are the same. In The Quanta Physics Theory we can do this.

A body at rest like the planets stars and galaxies to only name a few are all at rest. This is the subject of gravity when we review it. When a body begins to move or is upset for its rest sit, it is gravity that acts on this moving object.

An object at rest means it is not in contact with any exterior cause of motion. As you might know an object like a planet is at rest as it orbits and during rotation as we view it through a telescope. This orbitration motion is caused by space. In expanding space duration any and all celestials are in contact with this motion of the space. It can also be simply shown as in a cosmic drift. Even though the planets or celestial is in orbitration or moving it is caused by the cosmic drift. Relativity explains this space motion as an effect of space acting on the celestial at rest meaning that it not the action of the celestial in the independent sense of creating the motion.

The same events can be explained for planet mass objects. At rest they are not subject to external motion unless acted upon. When the mass or object begins momentum its internal mass-energy begins to be acted on. This interaction is what relativity conveys the increase and decrease of its energy mass. But this theory does not work in space. The reason is that space is an empty vacuum. A vacuum can be explained as an empty environment that when a celestial with life breathing deities are concealed due to the environment and loss of oxygen is mandated to maintain that life, is for whatever reason leaked to the vacuum the vacuum sucks all the interior air out of it.

So, in reference to that explanation the vacuum is empty all the time. On earth it isn't its concealed by the planets environment. Relativity doesn't work in space because of the exterior environment. Space is empty of energy negative or otherwise. For this reason relativity and the theory of gravitation is flawed. Accordingly, an all and any gravitation theory also known by Isaac Newton as 'universal gravitation' is essential.

On earth, the idea about gravity is governed by a negative and positive energy interaction that if it exists creates a division between space gravity and celestial planetary gravity theory. Though this is possible when we look at the universe as a whole we should understand that as a whole there should be a common connection between them.

If we view the gravity force on earth as a repulse force we observe a disconnection between its gravitation and the moving objects mass. In doing so we observe the idea that mass-energy is solely based on keeping a chunk of mass together in any of its form breaking the idea that an object in motion generates energy when it moves faster and faster on earth's surface is dis-connected as well. It is broken because of a law, that law states that any and all object masses no matter its size or energy of its mass subjected as 'similar" mass repel and do not attract as relativity claims.

In this case, an object at rest is dis-connected as being a connected energy mass. In essence, the Quanta Physics Theory works in reverse as relativity theory as such disconnected when in momentum the raise in its mass is subject to the energy to keep the mass intact as a solid mass and nothing else. If mass is subject to that velocity of light speed than the greater it is accelerated by an external propulsion it can only protect itself to the speed of light velocity. Though this can be debated by the empty vacuum and being disconnected to any external force acting on it - even as the ships velocity is increased beyond light speed having no force acting on it would confirm faster than light space travel.

On earth though, because of gravity that does act on moving masses relativity is correct that in a confined environment like the planet's atmosphere or any planet's atmosphere a limit may exist. Even though the atom bomb or later the nuclear bomb was a chain reaction of an explosive event and was not a singular event.

On earth as an object moves faster and faster it does 'not' gain weight or mass. It accelerates towards an escape velocity and not light speed because it doesn't need that much propulsion to enter into space and travel out of the earth's atmosphere. Light speed in this type of environment is a description of an absolute theory within an un-common boundary. The speed of light analogy should only be referred to in space flight outer space environmental conditions and not limited conditions.

The gravity force theories in non-gravity force conditions are unwarranted in a universal environment. The universe as a whole was created as a single super-repetition force. Relativity theory where the raising of a mass to infinity is disrupted as a propulsion force and not a objective mass stating that to try and push a ship no matter how much its weight, which is an exclusion of his own theory violates expressing the planet celestial gravitation force that has been measured with a mathematical number as a gravity force, where upon in outer space it has not and has been found to be un-conditional in any aftermath.

"Gravity Force only works when you take your foot-off the pedal and not just acceleration"

Gravity does not slow down object acceleration. Relativity has been misleading in this category about flight. Flight and acceleration is the product of just that acceleration meaning if I was to take my foot off the pedal (for acceleration) my ship, car and or plane will slow down and begin falling down from the air.

It is that the free-fall state measures at 9.8 m^e/second and my ship will fall relative to my ship's weight. As it falls I will feel acceleration left over from my last acceleration push force when I took my foot off the pedal. My speed will drop to zero and ZI will fall. My free-fall will continue until I reach the ground whether I glide in using the atmosphere's air or I crash like on " Lost in Space". My ship will go into free-fall until it reaches its full total weight and continue till it hits the ground either way or it comes to rest.

If I were to keep accelerating my velocity relative to my ships weight plus acceleration I could keep flying and reach escape velocity and reach space.

Traveling at the speed of light on earth at "c^1" my ships mass will square relative to my speed and weight, if I have more propulsion force and fuel than "c^1" I will travel faster than light relative to "g" (earth) because the mass of my ship or vessel will stabilize to "alpha-beta" and generate mobility mode. Mobility mode is when the alpha-beta waves in my ships mass stabilize and their frequency levels out to "c^2" (square). This will keep my ship and its body stable if I venture a straight path through the planet's atmosphere or in space. In space the stabilization occurs at a much lower frequency because there exist no known energy in the vacuum of space. As long as the ships manufacturing is micronized to an almost perfective micron degree everything should go well.

My ships mass (energy) like electricity when frequencies by acceleration has no mass whereas like 'light' that has no mass, inside a mass object 'alpha-beta waves will keep my ship together like it does a rock or stone on the earth's desert or ground. Like-wise electricity has no mass or weight when engaged by acceleration whereas propulsion influences my velocity. And as you know already 'light' has no mass in or outside a vacuumed space.

The Fundamental Laws of Gravity

$p=mv$

$9.8 \, m^e/sec. = 21.92 \, m.p.h.$

Which also equals: 1 mile a second

$9.8 \, m^e/sec = 1 \, mi/s$ (10 percent of 'c^1')

'c^1' = 15,534,429,2.233357

Laws of FTLST

Where light is constant is when: $E=mc = \frac{1}{2} E \, mc$?

On earth:

1) You can't travel pass light speed because (g) slows you down. Moving object will gain mass. $\frac{1}{2}^0 E=mc$

2) In space: you can't travel light speed because you gain mass - but no one can measure "g" in space because it's a vacuum. It's assumed to be a zero point vacuum, empty as its sucks up everything that is concealed inside anything like a tomato in a space suite is vacuumed out of it - because of faulty manufacturing or a tear. (terra)

3) Einstein's Biggest Blunder

Can we travel faster than light within earth's atmosphere "having the ability" to travel light speed.

'G' 'g' = $^2F=ma$ *quanta physics*

The free fall of earth's gravity, space and accelerations

The Isaac Newton verses Albert Einstein copyright delimena

Isaac's flawed lenses, Einstein's distortions

(g) = 9.8 m^e/sec or 1 mi/sec.

1 mi/sec per second (2,3,4 etc.) / 9.8 m^e.sec = 10% of 100 (100% is g).

1pd - 2 pds - 3 pds etc. = $\frac{1}{2}E=mc^1$, F=ma

Earth g = (p=mv) or $\frac{1}{2}$ E=mv. ? why? 1/10E=mv / (1/10 E=ma)

(g) = 1/10 of 1 mi/sec. (1/10 mi/sec. of 186,000 mi/sec.)

9.8 m^e/sec. = 1 mi/sec. = 186,000 m/s.

Can we travel faster than light stabilizing the energy in all matter

c^2 = 34,596,000,000 m/s.

Einstein $\frac{1}{2}$ E=ma,   Isaac Newton F=ma,   R. Kawecki ^{2}F=ma

c^1 = 186,000 m/s. or 670,616,629 m.p.h.

The Repulse of All Gravity

= pos + pos = repel.

Flaws in Relativity:

{mass = 186,000 m/s} a mass is 2 seconds., (93,000 m/2 sec.) 'g' fact. {earth}

You can travel faster than light inside a gravity uniformity gravity force if "G" is repulse. Two-specific a-like masses like the earth's mass (e) and a moving mass like a rock (mass (e).

The weight of all things:

You do not gain mass. {light mass (object + the mass @ (light-mass)

$c^1 \times c^1 = c^2 (2\ Lm)$?

c^2 = c squared. {Generating a stable mass - like moving electricity = weighs zero.

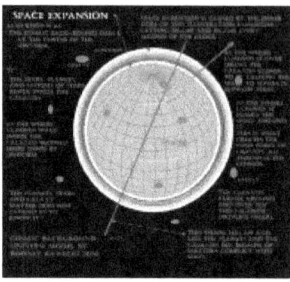

Space Expansion

Light is cancelled out by Darkness

Life is cancelled out by Death

These are our laws of Physics, and they like everything will change

FTL

Where velocity is greater than 'c' or the velocity of light

$v > c$ = 'λ' here is the de Broglie wavelength.

This is what validates v/c in relativity.

But electrons have zero mass whereas, matter particles do have mass, which is variable with gravity interference or planetary space volumes that begin at zero, and empty space where in vacuum matter, mass and weight realities diminish. v = wave..e = particle masses.

The reason energy and masses of material substrate can not increase or decrease energy polarity oscillations: What are Matter Waves?

Our definition for a wave has been too stringent – in What Is a Wave? We referred to (material) waves as oscillations of a medium about its equilibrium position in time and space. We introduced light as oscillations in the electromagnetic fields, but later discussed the validity of a particle model of light. Here, we still choose to refer to light as a wave because it obeys the principle of superposition. Superposition gives rise to constructive and destructive interference; an example of this is the two slit interference experiment.

Light of a given frequency, going through two narrow slits, superposes to give rise to the bright and dark bands {is the reference}

Matter particles, on the other hand, cannot interfere like this; they travel through slits one at a time.

So, it is by this conclusion that 'e' matter waves do not and cannot oscillate into increasing mass or decreasing the mass in matter material or what is the object.

What would electrons do if they were also passed through two slits? Because we are used to thinking of electrons as individual particles, we may expect a "particle-like" interference pattern like this:

When this experiment is actually performed, we get quite an unexpected result – the electrons form an interference pattern similar to waves:

The interference pattern created by the electrons is identical to one created by light with a wavelength but 'electrons" do not have a mass quantity so are wave-like.

AC vs DC

Duality 'e' wave polarity cannot increase without source

Therefore, dc matter is out and cannot act as a wave that polarates. It stays subtle and constant.

Whereas, a wave does not stay constant but is 'expandable'. It is not absolute.

Einstein matter wave hypothesis? Relativity,{matter acts as a wave but is physically defined as a particle and particles are not wave-like unless expressed in the medium therefore does not polarize on the account of 'e' is counter-intuitive to 'e' and repels as it's an absolute substrate matter material}?

What Enhances Space?

Like-wise a ship in space having *lightspeed* capacity will travel faster than c^1 because the vacuum rests the ship's matter material or vehicle shell's mass to zero points. The ship has no weight in space in the same way no sound, air or vibrations are recognized from the vacuum field.

Therefore, the ship having the ability has nothing there in the field to slow it down. In fact its propulsion forces of the ships engines increase because the vacuum field is physically being enhanced by the false vacuum that is constantly expanded and pushed by expanse. This expanse is a steady state of emissions that stream the expanse at second per second interval's engaging the expanse on the space vacuum.

The space vacuum has a ground state that is also constant. The extreme cold of space and vacuum absorption which also gives balance between the expanse volume and expanse velocity acts as a ground force that enhances the universe growth throughout space. Though, matter planetary matter does not expand the space and vacuum does.

This cycle of emissions aboard the cosmos brings forth balance throughout the universe. The more dense space inside the galaxies keep a steady state of orbital cycling and rotations amongst the stars systems in place as the expanse is put forth on the space that is pushing the galaxies away from one another.

The steady-state flow of the universe expansion based on Hubble equations transmits the over-all mechanism of the entire universe as a whole a refined obedient model of the universe's cosmos in the entirely a model of our universe in which perimeters are neither infinite nor finite in its entire. A new and unique idea about a universe that evolves over life-time evolutionary interval's from the big bang event that destroys one cycle but lays the foreground for the next. An endless cycle of planetary galactic life enhanced with its own biological life clock.

To understand this more clearly we should look at what causes mankind to a universal speed limit. The idea that without naming names was put on the universe by earlier information in physics. It broadcast the idea that matter and that's all matter gains mass or weight during accelerations. But the facts show that gravity or Newton Mechanics' illustrate an earth atmosphere of planet gravitation measuring 9.8 meters a second or one mile a second each second that correlates with as a method puts matter material at the brinks of zero point vacuum energy at zero meaning that 'weight' on a planet anywhere that has a gravity force unlike the moon is what defines an objects weight. Neil Armstrong proved this when we ventured the moon in 1963. As you already know because it was on all news channels that he dropped a feather right next to a iron hammer from his waist in front of him to test Einstein's analogy about 'weight' and the properties of mass {weight} and they both fell at the same time right next to each other together hitting the moon's surface parallel at the same time. This test proved not only that weight and mass are measured by the planets gravity force but that mass is what governed what we know as weight of objects. This universal weigh method of measuring masses throughout the cosmos places evidence in the vacuum of space that we know as a fact to measure a zero point vacuum quantized atmosphere. Where the entire universe is quantized by its variables weights and measures' the mechanics that define these forces have to interplay with each other or cancel out.

Gravity is one of these case analogies. So the universe ends with a bang but where after begins again with another transformation into another one. It acts on the common play of one event opening the gate into another era in time. No, it is as some people do believe about time travel correlations but with the idea that time travel is a closed subject when we understand the providence of inventing the factual ability to travel faster than the Einstein limit where the limit stops. Theoretically, the time has come and pass and has yet to make the critical information charts in space exploration. Engaging "*spacetime*" into the frame structure of an entirely governed commonwealth society in outer space where journeys are based on minutes and seconds instead of hours, days and years until then where amongst the planets and stars is yet a common place our gateway throughout the cosmos is our planets future directive until then.

The constant of logic at atleast twice light speed

Light is limited to its initial velocity 186,000 m/s, because it has no mass or very little everything of reality has mass even light. But mass something that has weight, having the ability to reach light velocity will travel faster when absent of gravity which gives it its weight because that energy is supported to acceleration.

On earth an object or particle mass travels relative to its weight which slows it down due to gravity. Without gravity such as in space vacuum the energy that gives mass its weight which how much energy it has, is converted to velocity. Since subatomic particles do not burn due to the lack of oxygen in the space the degree of collapse or diminish capacity does not exist at warp speeds.

For energy to have weight it needs a physical energy body like made up of protons, electrons, neutrons where photons are poverties of light not mass etc. Energy which are measurable in contest with light and accelerations contain both alpha waves and beta waves which inverted together by acceleration and exterior elements and forces of the environment combine acting parallel with gravitation energy is what gives mass its weight.

Without this combined set of forces or in a vacuum environment tat is empty of energy measured at zero points the energy of the objects mass is conserved. Without gravity the amount of energy of the body is sustained and converts the beta waves into what acted as a deterrent on alpha waves.

Momentum without the presents of gravity a mass body's energies explained as alpha and beta waves work together due to the lack of the third intermediate deterrent in the acceleration which is gravity these energy waves without the third deterrent work as particles when at rest. It is thought in 1906 in relativity that matter energies gain mass as acceleration increases so does its inertia energy but solely 99 percent of inertial mass is at the subatomic level negative mass or the orbitration of electrons in the state of inertia so in reality they do not gain mass but only orchestrate it sort of like expressing the key-rippling sequence on a piano.

```
        SUBJECT TO EXPANSION OVER TIME
           450,000 MILES PER SECOND ETC
       UNIVERSE ACCELERATING TIME LINE
           450,000 MILES PER SECOND
       INTERSTELLAR ACCELERATING TIME LINE
           440,000 MILES PER SECOND
       GALACTIC ACCELERATING TIME LINE
           25,000 MILES PER HOUR
       LOCAL EARTH ACCELERATING TIME LINE
       UNIVERSE ACCELERATION TIME LINE
```

Early modern physicists might not agree with the physics science analogy of the twentieth century but beta particles act like positive energies that in the attraction state attract but this is composed with the idea that in the nineteenth century the activity of gravitation was based on an objects rest state mass and that was defined as the reason moving objects were affected by gravity when they were moved so this attraction caused them to increase or decrease acceleration mass in the matter particles.

These are the laws of "similar a-like matter"

It was until the twenty-first century in 2020 that it was observed that no same like matter acted as an attraction state based on the common laws of physics called "similarity" Before this time it was thought that materials of the gravity state were based or due to 'attraction' between matter. Later it was discovered this attraction state was fore-given and was find to be wrong solely based on the properties of planet gravitation. Then, it was discovered that this analogy was mis-taken that no same a-like particles of matter or same material matter could attract that they repelled and were absolute as energy particle mass and these behaviors. It is a law that could not be changed and based on the idea that all material matter and or mass behaviors were a state defined in gravitation that the action preceded as an action between two same materials and therefore same quantities of the same material concept thus defined as positive properties. It was also defined that as such these same a-like materials would physically repel as followed the same activity between positive and negative electromagnetic behaviors thus such activity repel in the same manner as a positive and negative pole will attract two like poles deemed positive will repel.

The Nothingness of Space

The vacuum of outer space is not caused by the expansion of the universe, but is caused by gravity. First of all, when we say outer space (the space outside the atmosphere of planets and stars) is a "vacuum" or is "empty", we really mean that outer space is nearly empty or almost a perfect vacuum.

A vacuum is an empty place, which space nearly achieves. ... But the vacuum of space is the opposite. By definition, a vacuum is devoid of matter. Space is almost an absolute vacuum, not because of suction but because it's nearly empty.

Quanta Physics scientists believe the vacuum or the quantity of empty space is quantized. That at the time of the big bang is the event that also quantized all matter existing. Where vacuum fluxuations are much smaller than matter particles when the bang happen space quantized the explosion into its smallest molecules. In the explsion space fluxucations stretched in the suddeness forming these small fluxuations into a flat fabricated yet quantized flexible one dimensional surface.

A one dimensional surface by which length, size and depth were acceleratable to this fabric.

Such surface quantities of quantized fluxuation fibers generated from the smallest of vacuum bottom particle vacuum molecules measurable at zero point that generate at masses smaller than quant particles and behaviors, is enought to quantize everything in the vacuum space.

Quantize matter without mass was regenerated during the first bang of the universe simulation that vacuum quantizing matter into mass creating matter with weight that later is generated by spectrum gravitation anomalies that were spread throughout the cosmos. The deep space uniformity of a preival atom universe directed by a single arrow of time expansion line expresses the expanse with a diminish capacity over time when united into a single event of cosmology.

Hubble expanse with Kawecki space expansion as an unknown yet unknown anomaly that isnt defined until 2020 CY, defining the assumed natural of the expansion caused by later affects that are secondary to the big bang event evolution - space expansion by means of an un-natural sphere anomaly hidden underneath the void of space this inflationary idea resumes expanse as an anomaly preceded by an inflating sphere also recognised in the cosmic back ground imagery. A gigantic expanding sphere expanding the matter universe through the inflation of a cosmic sphere existing invisible to the known universe perimeters.

As we travel into orbit, outer space and ultimately intergalactic space, the pressure varies by several orders of magnitude. Atmospheric pressure is variable but standardized at 101.325 kPa (760 Torr).

Two objects exert a force of attraction on one another is known as "gravity." Sir Isaac Newton quantified the gravity between two objects when he formulated his three laws of motion. The force tugging between two bodies depends on how massive each one is and how far apart the two lie. This description asserts that even as the center of the Earth is pulling you toward it (keeping you firmly lodged on the ground), your center of mass is pulling back at the Earth. But the more massive body barely feels the tug from you, while with your much smaller mass you find yourself firmly rooted thanks to that same force. Yet Newton's laws assume that gravity is an innate force of an object that can act over a distance.

Einstein realized that massive objects caused a distortion in space-time. Imagine setting a large body in the center of a trampoline. The body would press down into the fabric, causing it to dimple. A marble rolled around the edge would spiral inward toward the body, pulled in much the same way that the gravity of a planet pulls at rocks in space.

But this didn't explain whether space acted in a single one dimension or is a three dimensional body?

At Gravity Falls

So, does this really mean there exist a force of attraction from the planets center of mass?

No.

Why because the link of attraction is broken by the falling of gravity. If things didnt just fall measured at earth 9.8 meters a second, as on the moon they fall slower. It shows that the exist no connection between the planets mass and falling objects. Yet this is the interpetation of relativity.

In space though, things dont just fall - they sit still. Or as still as one would think since they are not moving but they are. Space pushes against material matter objects like objects, planets and stars shows that space has a quantity assertion of massive objects and that force is asserted on matter objects in space. So, the question is whether the space and its vacuum cause matter to spin and or rotate or stop things from just falling and bend in the presents of matter?

The answer is either. The force of the vacuum space is the

torque of its gravity as we understand it. The force puts planets, stars and the galaxies into motion. It is this action that creates what relativity asserts as a trampoline.

The actual asertion of the vacuum force is stillness. Expansion asserts the vacuum into motion because the universe is expabding. Expansion makes the universe in motion from otherwise keeping still in the nature of its inhabitence. Expansion started out slowly whereas over time emerged into a critical emergence. The expanse clause was created secondary to the universe's creation where upon in that creation a hiden entity evolved slowly changing the stillness or calm afer the bang and it has been expanding ever since.

Where do we stand with Warp Drive in the Twenteith Century

Warp drive in Star Trek works by annihilating matter (in the form of deuterium, a kind of hydrogen gas) and antimatter in a fusion reaction mediated by dilithium crystals. This produces the enormous power required to warp space-time and drive the ship faster than light.

Is warp drive possible?

Warp Drive May Be More feasible than thought, scientists say. A ring-shaped warp drive device could transport a football-shape starship (center) to effective speeds faster than light. The concept was first proposed by Mexican physicist Miguel Alcubierre.

What is a warp drive engine?

A warp drive is a theoretical superluminal spacecraft propulsion system in many science fiction works, most notably Star Trek and much of Isaac Asimov's work. A spacecraft equipped with a warp drive may travel at speeds greater than that of light by many orders of magnitude.

How fast is warp 1 in mph?

The speed of light in a vacuum is about 186,282 miles per second (299,792 kilometers per second). In "Star Trek," a warp factor of 1 is light speed, and a warp factor of 9.9 is more than 2,000 times greater than light speed.

FTL in Numbers

All faster than light space travel acceleration analogies stem from relativities acquiring $e = m\,'C'$ square.

'C' sqaure is equal to about 34,000,000,000,000 or 34 billion atomic energy units or electron volts per atom.

Divide this 'c' square by the speed of light and your answer will equal hyper drive.

There are 186,000 warp drives in 'c' square.

'C' squared which is used to count atoms in a chunk of mass, divided by its length in 3D, will equal its mass.

But accelerations are many when applying these numbers to velocity.

So, science fiction like star trek movies use the velocity of light in dividiond of 100 per centages to coordinate a velocity in space travel.

How fast can we go in space?

About 16,150 mph, thats equal to about how fast one of the united states fastest jet travels on earth but in space 10 percent is added. Only ten percentage is added because thats the fastest the countries propulsion engines are limited too. Make sense?

1996 Out into space

Once at a steady cruising speed about at (26,000kph) in orbit, astronauts no more feel their speed than do passengers on a commercial airplane.

Earth Atmosphere

1. thrust
2. lift

3. drag

4. weight

Space

1. thrust
2. drag

So, if space is an empty vacuum how do we get thrust?

By force. The space vacuum is quantized to the vacuum manifestion of waves. This means that any un-firmilar action of matter objects taking action in space - the vacuum protects itself by quantizing a resistents.This resistents is equal to the force implied by the object that wishes to accelerate or move in space.

Space therefore, is three-demensional and scaled to the vacuum resistents scale. For that scale the vacuum measures close to zero or is almost a perfect vacuum having no matter, particle or gasious molecules in it.

There are some elastistic particles that are un-attached from matter energies like the suns and flamious planetary matter that exist throughout space. But these particles are just that small particles that have no dividen to accumulated chunks of matter.

Space is the calm of a vacuum or empty space but is uniform.

The manifesation of the space vacuum is zero point and almost perfect yet instantaneious quantization of the vacuum that accumulates at aany point of changing the calm of the vacuum collective or making of waves. It manifests and accelerates when the natural stillness of the vacuum is desturbed. Since the vacuum is the smallest alignment of visible anomaly as for its consistency to something that will create disturbance waves in its enviornment everything that exist in the universe is confined to instant quantization of repulsive space fluctualtions and flexibility existing in the empty vacuum.

No lift

No drag

No weight

Matter making Vacuum Waves

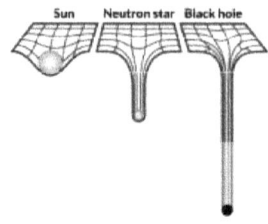

www.ingramcontent.com/pod-product-compliance
Lightning Source LLC
Chambersburg PA
CBHW060844220526
45466CB00003B/1228